The Forts of Old San Juan

San Juan National Historic Site
Puerto Rico

Produced by the
Division of Publications
National Park Service

U.S. Department of the Interior
Washington, D.C.

National Park Handbooks are published to support the National Park Service's management programs and to promote understanding and enjoyment of the more than 370 National Park System sites that represent important examples of our country's natural and cultural inheritance. Each handbook is intended to be informative reading and a useful guide before, during, and after a park visit. They are sold at parks and can be purchased by mail from the Superintendent of Documents, U.S. Government Printing Office, Washington, DC 20402-9325.

Library of Congress Cataloging-in-Publication Data
The forts of old San Juan: San Juan National Historic Site, Puerto Rico/produced by the Division of Publications, National Park Service.
  p. cm.—(National park handbook series; 151)
  1. San Juan National Historic Site (San Juan, P.R.) 2. Fortification—Puerto Rico—San Juan—History. 3. Puerto Rico—History—To 1898. 4. San Juan (P.R.)—Buildings, structures, etc. I. United States. National Park Service. Division of Publications. II. Series: Handbook (United States. National Park Service. Division of Publications); 151.
F1981.S2F6 1996 623'.197295—dc20 96-21604 CIP
ISBN 0-912627-62-X

☆GPO 2006:320-367/20007 Reprint 2006
Printed on recycled paper

For sale by the Superintendent of Documents, U.S. Government Publishing Office
Internet: bookstore.gpo.gov Phone: toll free (866) 512-1800; DC area (202) 512-1800
Fax: (202) 512-2104 Mail: Stop IDCC, Washington, DC 20402-0001

ISBN 978-0-91-262762-5

Contents

Part 1 **Old Forts in a Modern World 4**

Part 2 **Gateway to the Indies 14**
Strategic Seaport 18
Point of Entry: The San Juan Gate 20
La Fortaleza 26
1595: Drake's Attack 32
1598: Cumberland's Attack 36
Ordnance 40
How a Siege Works, Circa 1700 42
1625: The Dutch Attack 44

Part 3 **A "Defense of the First Order" 50**
El Morro 54
Aspects of a Soldier's Day 56
San Cristóbal 62
Of Cisterns and Bulwarks 64
1797: Abercromby's Attack 70
San Juan Under Fire 1898 74
Defenders of Old San Juan 76

Part 1

Old Forts in a Modern World

The present entrance or sally port to El Morro is reached by an arched masonry causeway that dates from the late-18th century. It replaced a wooden drawbridge.

Preceding pages: *El Morro dominates the harbor entrance like a mighty battleship. The fortress, along with other fortifications around the city, made 18th century San Juan one of the best-protected ports in the Caribbean.*

Old San Juan lies at the gateway to the Caribbean from Europe. It is situated on the western end of a small barrier island that lies between the broad San Juan Bay and the open sea. A steep headland overlooks the only navigable entrance to the town's deep and landlocked harbor, and the island's rocky seacoast is skirted by a treacherous reef over which long Atlantic rollers crash like thunder. Boquerón Inlet, at the eastern end of the island, is too rocky and shallow for large ships and forms a natural water barrier to the harbor.

In the days when sailing vessels followed the northeast trade winds that blew from Africa to the West Indies, the port was a safe haven from tropical storms. But far more important, it provided a secure naval base from which to control shipping into the Caribbean and on to the shores of Mexico and South America. The English, and later the French and the Dutch, wanted the port for this very purpose. Consequently, although Puerto Rico was an ocean away, Spain was obliged to hold it firmly or risk the loss of its other territories in the Americas. The fortifications of San Juan cost Spain dearly; but because of them, would-be attackers faced almost insoluble logistical and strategic problems.

In fortifying the old city, Spanish military engineers made maximum use of San Juan's natural defensive features and erected fortifications that controlled both land and sea approaches. Many of the works also served as operational bases for Spanish military and naval units.

The Spanish forts in the city of Old San Juan have evolved in a surprisingly logical way from their 16th-century nucleus, much of which still exists. Part of their construction also dates from the 17th and first half of the 18th centuries, but most of the massive fortifications seen today were built between 1765 and 1800. They reflect the best military thought of their time.

The story of those fortifications and how they grew is a tale that mirrors the history of great world powers. It is also a tale of gold and silver, pirates, sugar, and slaves, and the impact of the trade winds on the history of this sun-drenched tropical isle.

Right: *Old San Juan was once completely enclosed by a wall that linked the major fortifications and helped to protect the city against attacks from any direction. Portions of the wall were removed in the 1890s to allow for the city's expansion*

This small, 50-foot-square fort stands across from El Morro on the west side of the harbor entrance. Its official name is **San Juan de la Cruz,** *or "St. John of the Cross," but most people today call it El Cañuelo after the tiny island on which it was built in the early 1660s. Its walls stand about 15 feet high. The flat roof provided a platform for cannon. Beneath this deck are the ruins of a cistern and powder magazines. El Cañuelo was situated so that its fire sector overlapped El Morro's. The little fort also defended the mouth of the Bayamón River, which linked San Juan with the backcountry. El Cañuelo was far more than a simple artillery battery. It was a closed redoubt with the magazines, quarters, and cisterns to sustain a garrison for many days.*

Next pages: *The courtyard or main plaza of San Cristóbal, a massive fortification whose completion in the early 1790s gave San Juan a "Defense of the First Order." Its architecture reflects the engineering competence and craftsmanship, as well as the cost, that went into creating a first-class fortification.*

Left: *La Fortaleza was the first "fort" built to protect San Juan and its harbor. It has been the traditional residence of Puerto Rico's governors since the late 1500s. It is the oldest executive mansion still used as such in the western hemisphere.*

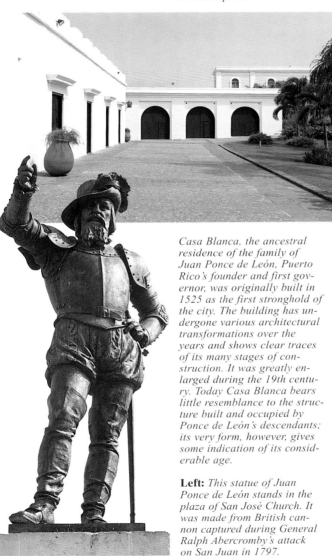

Casa Blanca, the ancestral residence of the family of Juan Ponce de León, Puerto Rico's founder and first governor, was originally built in 1525 as the first stronghold of the city. The building has undergone various architectural transformations over the years and shows clear traces of its many stages of construction. It was greatly enlarged during the 19th century. Today Casa Blanca bears little resemblance to the structure built and occupied by Ponce de León's descendants; its very form, however, gives some indication of its considerable age.

Left: *This statue of Juan Ponce de León stands in the plaza of San José Church. It was made from British cannon captured during General Ralph Abercromby's attack on San Juan in 1797.*

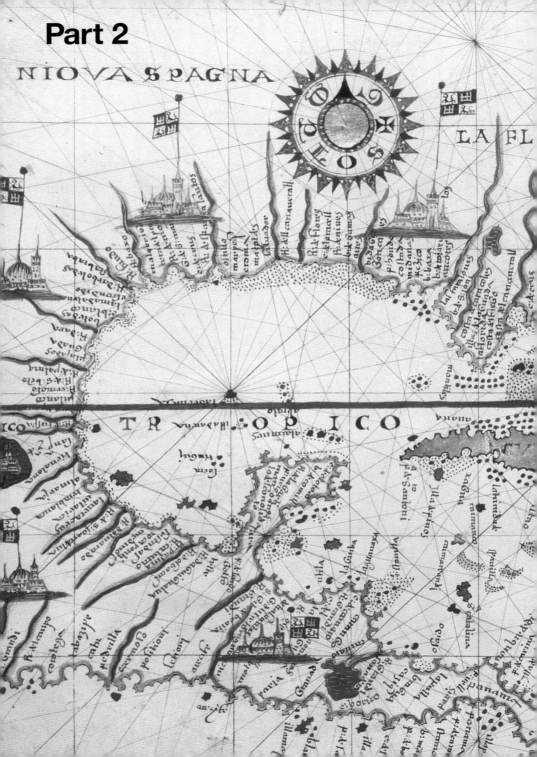

Gateway to the Indies

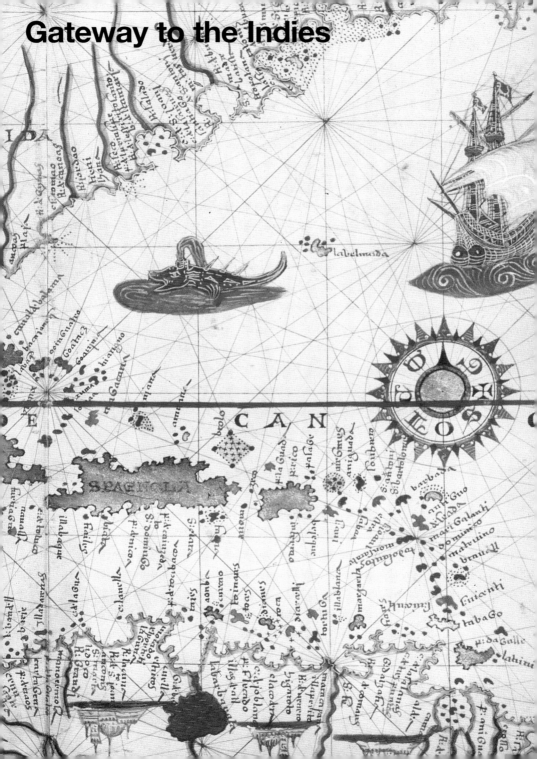

A curving chain of islands stretching from Florida to South America forms the eastern edge of the Caribbean Sea. Collectively these islands are known as the West Indies. The upper half of the chain is divided into two groups: the Bahamas, comprising a large number of small islands, cays, and reefs; and the Greater Antilles, which includes the islands of Cuba, Jamaica, Hispaniola (now Haiti and the Dominican Republic), and Puerto Rico. The remainder, another large group of small islands, among them Martinique, Trinidad, and the Virgin Islands, are known as the Lesser Antilles.

During the 16th century, Spain used the Greater Antilles as a base for her attempts to explore, settle, and exploit the resources of Mexico and Central and South America. By 1535, with the conquest of Peru and Mexico, Spanish ships were voyaging homeward laden with gold, silver, and jewels. To safeguard these "treasure fleets" and her New World discoveries against pirates and traditional enemies like England, Holland, and France, Spain instituted a convoy system that compelled most ships sailing in the American trade to travel under the protection of armed vessels. She also began to build mighty fortifications at key points along the fleets' routes in the Gulf of Mexico and the Caribbean Sea. One such key point was on the island of Puerto Rico, at the eastern edge of the Caribbean.

Puerto Rico had come to European attention on 19 November 1493, during Christopher Columbus's second voyage to the New World. At that time the island was home to several thousand Taíno Indians of the Arawak culture who called the place *Boriquén*, meaning Land of the Brave Lord. Columbus named it "San Juan Bautista" (after St. John the Baptist), but it was Juan Ponce de León, explorer of Florida and the first governor of Puerto Rico, who provided the name by which the island is known today.

Ponce de León first came to Puerto Rico with Columbus in 1493. He stayed only a brief time, however, before sailing on to Hispaniola (then called Santo Domingo), which he helped to settle, defend, and govern for the next 15 years. His role in subjugating the Indians in Hispaniola's eastern provinces won the attention and praise of Don Fray Nicolás de Ovando, Royal Governor of the lands claimed by Columbus. Thus when the time came to explore and

The Puerto Rican coat of arms dates from 1511 and is one of the oldest coats of arms still in use in the hemisphere today. The lamb symbolizes St. John the Baptist, patron saint of the island. The castles, lions, flags, and crosses represent the different kingdoms claimed or governed by Spain's Catholic rulers. The letters "F" and "Y" with the yoke and arrows are personal emblems of Ferdinand and Isabella, the ruling sovereigns when Christopher Columbus first came upon the island.

Previous pages: *Spain's New World empire, from a somewhat fanciful 16th century Spanish map.*

colonize Puerto Rico, Ovando turned to Ponce de León to carry it out.

Ponce de León reached the island with a troop of 50 soldiers on 12 August 1508. They soon found the harbor we know today as San Juan, and he called it a *puerto rico*, a fine or excellent port. (Curiously, in later years the harbor and the island traded names: the island became Puerto Rico and the port, San Juan. Boriquén, the Indian name for the island, was forgotten.) After some indecision, Ponce de León settled his people not on the fine harbor but about two miles south at a wooded site surrounded by hills and swamps. He named the settlement Caparra after an old Roman village in Spain.

Despite odds, this determined group clung to their foothold. Some began to clear the lands that one day would become broad plantation fields; others sought the elusive gold that had helped lure them to the island in the first place. In the early days, the native Taínos were friendly and helpful. But friendship turned to hostility as more Spaniards arrived and robbed the Indians of their lands and their women, and forced them to mine gold and till the fields in the manner of slaves.

In 1511, the Taínos rebelled against their oppressors. While they fought with great courage and determination, their primitive bows and arrows, stone axes, and wooden swords were no match for Spanish firearms. After the rebellion was crushed, many of the Taínos fled to the Lesser Antilles where they joined forces with the Island-Caribs, a fierce tribe of South American Indians who hitherto had been their enemies. Together they began a campaign of terror and harassment that menaced Puerto Rican settlements for nearly a quarter century.

Caparra turned out to be a less-than-ideal site as a military base and seat of government. The surrounding swamps made the location unhealthful and inconvenient. It was also hard to defend and far from the sea. The colonists urged Ponce de León, who had been named governor of the island, to move the settlement, but he refused. The Spanish government, however, overruled him.

For their new settlement, the colonists chose the islet of San Juan, "the best and most beautiful site in the world." The little island was only three and one-half miles long and about a mile wide at its western

Strategic Seaport

Like several other Spanish ports in the Caribbean, San Juan, shown below as it appeared in the early 1790s, played a strategic rather than a commercial role in Spain's imperial scheme of things. It was not a major link in the convoy system, and the great treasure fleets did not stop there except to repair damaged vessels or for other emergencies. The little commerce the port enjoyed with the outside world was

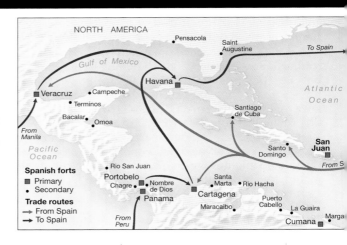

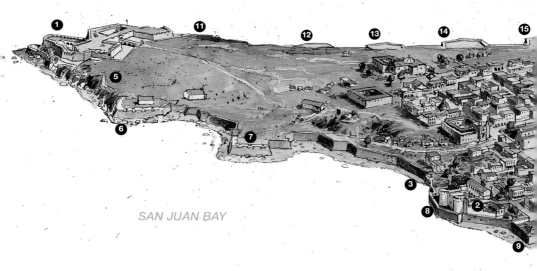

1	El Morro	9	La Concepcion Bastion	17	La Puntilla
2	La Fortaleza	10	San Jose Bastion	18	San Justo Bastion
3	San Juan Gate	11	San Antonio Bastion	19	San Justo Gate
4	San Cristóbal	12	Santa Rosa Bastion	20	Muelle Bastion
5	San Fernando Bastion	13	Santo Domingo Bastion	21	San Pedro Bastion
6	Santa Elena Bastion	14	Las Animas Bastion	22	Santiago Bastion
7	San Agustín Bastion	15	Santo Tomas Bastion	23	Santiago Gate
8	Santa Catalina Bastion	16	San Sebastían Bastion		

mainly indirect through Havana and Santo Domingo and confined to a considerable but relatively minor trade in sugar, coffee, hides, ginger, and other tropical products.

Puerto Rico's geographic position at the eastern edge of the Caribbean made San Juan one of the key frontier outposts of Spain's West Indies dominions. The forts here and those in Cartagena, Portobelo, Havana, Veracruz, and St. Augustine formed a chain of defenses that guarded Spain's New World trade routes (shown on the map at left) and helped to protect the treasure ships that vitalized her empire.

King Philip IV called Puerto Rico "the front and vanguard of all my West Indies . . . the most important of them all and the most coveted by my enemies." The San Juan fortifications served to keep the island and its excellent port out of the hands of Spain's enemies, notably England, Holland, and France, who by the middle of the 17th century had occupied the Lesser Antilles to the southeastward. In their hands, San Juan and Puerto Rico would have provided a base for raids upon Spanish trade and settlements.

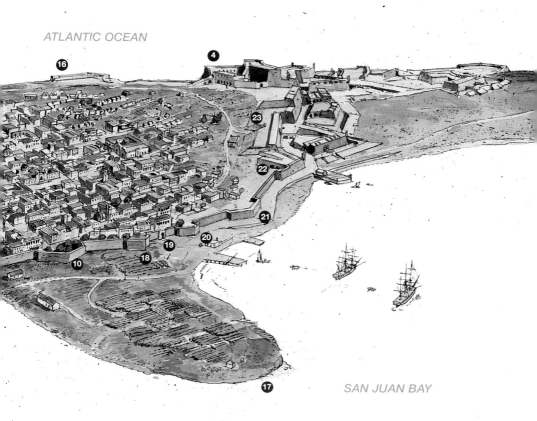

Point of Entry: The San Juan Gate

Nothing evokes the historic sense that Old San Juan was seen by its founders and residents as a *puerto rico* or "excellent port" better than San Juan Gate at the foot of Caleta de San Juan, the street leading down from the Cathedral de San Juan Bautista. It is the oldest of the three original gates that once gave access to the city and the only one still standing. The other two gates were demolished in the 1890s when portions of the east and south walls were removed.

San Juan Gate, shown here as it might have looked in the mid-17th century, was built by Governor Don Enríque Enriquez de Sotomayor in 1635. It fronts a quiet little cove that, in the early years, was the anchoring place for ships in port. Here crews unloaded cargoes from fast lateen-sail rigged dispatch boats, common merchant ships, and lumbering galleons, those towering seagoing wooden fortresses that represented the formidable seapower of imperial Spain. The mariners, officers and crew alike, having survived storms and pirates and other dangers of the deep, usually made the Cathedral their first stop, to give thanks to God for their safe arrival.

For a long time, San Juan Gate was the main entrance to the city from the harbor side. Through its massive *ausubo* doors passed the money to pay the garrison and build the fortifications, the *situado* brought from the viceroy's treasury in distant Mexico City. Through this gate also came the voluminous correspondence, the instructions, reports, and inquiries, as well as the men who were expected to administer the king's will and justice in these lands.

Following the traditions of walled cities from Biblical times, it was here that new arrivals sought permission to enter from the civil authorities. (These officials were close at hand, the governor's residence, La Fortaleza, being only a few dozen paces away.) It was here, too, that official ceremonies welcoming new governors and bishops were conducted before they entered the city to take up their duties. These were usually occasions for much celebration, in which the whole city took part.

end. It lay just at the harbor entrance and was open to cooling winds from the Atlantic. The terrain was naturally defensible: jagged reefs lined the Atlantic escarpment, and the shoreline on the harbor side near the village was craggy and steep.

The transfer began in 1519 and was completed in 1521, the year Ponce de León left to colonize Florida. He never returned to Puerto Rico. In Florida, where he tried to establish a settlement, Indians drove off his ships and fatally wounded him. His remains were later brought to San Juan, and are now entombed in the Cathedral de San Juan Bautista.

San Juan was a superior location—the king of Spain had even called it the "Key to the Indies"—but the area possessed few natural resources and little of the fabulous wealth that men were finding in South America and other parts of the Caribbean. The gold that had helped draw the Spanish to Puerto Rico in the first place petered out after 1540. Agriculture flourished, but there were never enough slaves to make farming productive. And the plantations were vulnerable to Carib attacks from the Lesser Antilles.

Indians, however, were only part of the threat confronting the new colony. What troubled Spanish officials even more was the coming of French corsairs to the Caribbean. Animosity between Spain and France kept the two nations almost continually at war from 1520 to 1556. After 1522, when a French privateer first showed his ambitious king, Francis I, American treasure taken from Spanish vessels on the high seas, French mariners searched avidly for Spain's New World commerce. Many French corsairs left their customary hunting grounds off the Spanish coast and sailed west to the Indies where the chances for plunder, both at sea and ashore, were vastly greater and where Spanish settlers had no protection except locally improvised militia.

In 1528 the French sacked and burned San Germán on Puerto Rico's west coast, and for many years thereafter used island waters almost at will. Among the most notorious of the raiders was François le Clerc, called *Pata de Palo* ("Peg Leg"), who knew the coasts of Puerto Rico and Hispaniola like the palm of his hand. Jacques Sores was another. With but a single vessel and a hundred archers, he sacked at least five Caribbean towns. It was rumored that his patrons included Queen Elizabeth of England.

Don Pedro Menéndez de Avilés, famed seaman and soldier, sometime Captain General of the Fleet of the Indies and the colonizer of Florida. He helped develop the convoy system to protect the Spanish treasure fleets in their annual voyages from the Americas to Spain.

The coming of European enemies to the Caribbean added a new dimension to Spain's defense problems. Corsairs off the coast of Spain had been a hazard for generations—a challenge to be run, a cruel and ofttimes deadly experience for young Spaniards starting naval careers in the king's service. But Spanish ports were strongly fortified, the *guarda costa* (Spain's Coast Guard) was active, and important shipments were invariably protected by convoys. In the Caribbean, however, there were no strong forts and no *guarda costa*. Even if the islands had no great wealth, they served well as bases from which corsairs could cruise against Spanish shipping, or as depots for commerce or smuggling.

To counteract her mounting shipping losses in the Caribbean, Spain introduced that ancient naval device, the *Armada de la Guardia* (the convoy system), under which most ships involved in the American trade were required to sail under the protection of armed vessels. The system involved two fleets. One, called the *flota*, left Spain in the spring for the Gulf of Mexico, dropping off merchantmen at Puerto Rico, Hispaniola, and Cuba, and making port at Veracruz, Mexico. The second fleet, the *galeones*, departed Spain in midsummer for Cartagena on Colombia's northwestern coast and Portobello at the Isthmus of Panama. With the *flota* carrying trade goods and Mexican silver and the *galeones* transporting silver, gold, pearls, and other precious stones, the two fleets would meet at Havana the following spring and sail together back to Spain, exiting the area through the Florida straits. In this fashion, the convoys brought supplies and merchandise from Spain and took back annual cargoes of treasure and raw commodities from the colonies.

In a further move to protect her New World holdings, Spain declared the Caribbean a *mare clausum*—a closed sea. All shipping had to be in Spanish bottoms and foreign vessels were subject to seizure. But this decree assumed that Spain controlled the seas, which in truth she did not. Blasco Núñez de Vela, who led the first convoy into the Caribbean in 1537, told King Charles V that the navy could handle the corsairs and that there was no need for forts except minor defenses against the Indians. The crown's advisers, however, urged the king to fortify the key ports of the Caribbean, including the islands.

In those days Puerto Rico had no fortifications to deter an invader or to serve as refuge in time of danger. There was only Casa Blanca, a substantial structure of tamped earth and stone owned by the Ponce de León family and licensed by the Spanish crown for use as an arsenal for weapons and a repository for government funds. Strategists pointed out often and loudly to the king that Puerto Rico was "the entrance and key to all the Indies...the first to meet the French and English corsairs.... Your Majesty should order a fort built...or the island will be deserted."

The crown authorized a permanent fort at San Juan on 30 May 1529, but Indian problems delayed construction until 1537. This fort, which came to be known as La Fortaleza, was built near the rocky shore of the bay, not far from Casa Blanca. Construction was almost completed by 1540, but without cannons or garrison (there being no money for guns or soldiers), the new building was almost useless for military purposes. Actually San Juan officials had asked for very little: only six men, including two gunners, two guards, a man to keep the weapons in condition, and a porter. They wanted a half-dozen 8-pounders, 20 arquebuses, 20 crossbows, and 40 pikes. Some of this materiel was on hand by 1542, but the long-range guns, so badly needed for harbor control, were interminably delayed.

From the beginning, many observers complained that the location of La Fortaleza was a mistake. The noted Spanish historian, Gonzalo Fernández de Oviedo, who saw La Fortaleza when construction started in 1537, reported that "only blind men could have chosen such a site for a fort." Although it overlooked the anchorage and controlled access from the harbor to the town, La Fortaleza had no command of the harbor entrance. Since it could only be seen when inside the harbor, an enemy would naturally think the town was unprotected. The fort should have been built, said Oviedo, as lofty as a watch tower and on *el morro*—the headland—at the port entrance where the land slopes precipitously down more than 100 feet to the sea. A few cannon there, he and others argued, would lock the harbor against any kind of hostile naval craft.

Oviedo's criticism seems to have had some effect because within two years the crown approved funds to fortify the headland. Against the cliff midway up the slope (some 60 feet above sea level) Spanish workmen built a vaulted masonry tower with embrasures for four cannon. Three cannon were initially mounted here, but every time they were fired the interior filled with smoke that blinded and choked the gunners. The guns were soon moved outside. A semicircular platform for three guns, connected to the tower by a series of steps, was also constructed over the rocks at the foot of the slope. This came to be known as the Water Battery. Smartly handled, the low-level guns mounted here could sink incoming enemy vessels with waterline shots.

After the tower and the Water Battery were completed, military affairs played a limited role in the development of San Juan until 1582, when Captain General Diego Menéndez de Valdés came to govern Puerto Rico on behalf of Philip II, who had ascended to the throne of Spain on the abdication of his father, Charles V, in 1556. The new governor was a veteran of the Florida campaigns and no stranger to the West Indies, nor to frontline military posts. He was in San Juan because Philip II had decided to make the town a presidio, with the 50 soldiers of the captain's company its first garrison. Under the presidio system, Menéndez de Valdés became the first governor with the title of Captain General of Puerto Rico. Within 12 months he had prepared a no-nonsense report on this Caribbean gateway that eventually reached and impressed the high councils of the Crown.

In all the Indies, said the governor, the harbor at San Juan was the best and easiest place to fortify. And if it ever got into enemy hands, it would be impossible to recover. His recommendations (build forts at key harbors and establish a strong coast guard) were neither new nor extraordinary. But this time the report came to the right people at the right time. And because its impact went beyond Puerto Rico into all the Caribbean, the king's counselors formed a committee (the "Board of Puerto Rico"), which evolved into the *Junta de Guerra de Indias* (the War Council), the powerful and permanent arm of the Crown that handled all matters concerning the defense of the West Indies.

Menéndez de Valdés, while awaiting official action on his report, worked out an interim plan and began to implement it by erecting palisaded earthworks at vulnerable landing spots around the islet. Perhaps

La Fortaleza

La Fortaleza was the first fort built to protect the harbor and city of San Juan. By European standards it was not much of a fort, but, of course, it was intended principally as a defense against a Stone Age people—the Island-Carib Indians. On the land side it looked like just another flat-roofed house. But its walls were seven feet thick and in a pinch could shelter a couple of hundred people. While it overlooked the anchorage and controlled access to the town from the harbor, La Fortaleza had no command of the harbor entrance—a factor that brought its effectiveness as a defense work into open criticism.

The main gate faced the town and was protected in front by a small crescent-shaped defense work called a demilune. The area between the house and the shore was enclosed by a wall six and one-half feet high and embrasured for cannon. A tower at one angle of the wall afforded not only a vantage

point for the defenders but a "place of homage" where the governor or warden took his oath of loyalty to the Crown. Later a second tower was built. Each had a strong, vaulted room—one was used for the storage of gunpowder, the other as a prison.

La Fortaleza has been the residence of Puerto Rico's governors since 1544. Other Royal officials lived here, too, and it was the repository for treasury documents and the Chest of Three Keys, which held the funds to finance the colony. It also acquired an elegant new name: *El Palacio de Santa Catalina.*

Except for the towers and the main walls, La Fortaleza was destroyed by the Dutch in the 1625 attack on San Juan. It was reconstructed in 1639-40. Although the structure has been added to over the years, the two 16th-century towers are among the earliest examples of military architecture in the Americas.

the most important thing he did was to build a four-gun battery called Santa Elena at a high point along the shore between El Morro and La Fortaleza. This battery, located near a promontory called *Cerro de los Ahorcados* (Hangman's Hill) within musket shot of the harbor channel, would prove its worth when Sir Francis Drake attacked San Juan in 1595; eventually it became the strong Bastion of Santa Elena that stands today.

Like other struggling Spanish settlements in strategic locations from San Juan to St. Augustine, Puerto Rico did not have the resources to carry out its military mission alone. But King Philip II decreed that the Caribbean *must* be fortified and that the viceroy of New Spain (colonial Mexico) must pay the bill. From the Crown's viewpoint, this was sound logic since each of the proposed new defenses was designed to protect New Spain and its shipping. Administratively, the boundaries of New Spain extended from Central America and the Antilles to Canada. Today's Mexico was its heart—a heart of silver from the fabulous mines. Philip's policy meant that each defense community on his list was eligible for annual funds to guarantee its maintenance. These *situados* (subsidies) served as the fiscal foundation for San Juan and other Caribbean military towns until the end of the 18th century.

Although the arrival of Menéndez de Valdés and his men gave San Juan a garrison of professional soldiers for the first time, their presence did not mean that Puerto Rico was now a safe place to live. With only 170 poverty-stricken families, San Juan actually had fewer inhabitants than at its founding six decades earlier. The rest of the island was but thinly populated and had urgent defense problems of its own. For 40 years corsairs had used the island's ports almost at will, plundering as they chose and jeering at puny local militia who knew that a determined enemy could sweep them off the island.

Sir Francis Drake was such an enemy. Few men knew Caribbean waters, ports, and people as well as this personification of England's emerging seapower. Drake had been actively challenging Spanish shipping in the Caribbean since 1567, when he was a young captain in the fleet of his cousin, the slave trader and merchant John Hawkins. Whether one thought of him as a pirate or knight, Sir Francis

Drake was the English counterpart to the Spanish conquistador; both had the skill, courage, and leadership that drove them to incredible deeds.

The exploits of Drake and other English freebooters in the Caribbean in the 1580s underscored the danger Puerto Rico faced. The unrestricted landings and pillaging and the easy capture of Spanish ships also showed Spain's vulnerability. Philip II reacted to these threats by sending two experts into the Spanish West Indies to plan the defenses needed to preserve his overseas domain: Field Marshal Juan de Tejeda, a military veteran bearing the scars of many battles, and Juan Bautista Antonelli, a talented Italian engineer-architect.

Tejeda and Antonelli reached the Caribbean in 1586. By the time they returned to Spain in September 1587, they had inspected every major port from the Antilles to the Spanish Main, including Florida. The plan they finally submitted to Philip proposed to fortify ten key sites: San Juan in Puerto Rico and Santo Domingo in Hispaniola; Santa Marta and Cartagena in Colombia; Nombre de Dios, Portobelo, the Chagre River, and Panama City in Panama; Havana in Cuba; and St. Augustine in Florida.

This plan, which the king authorized in November 1588, had matured at a critical time. England's victory over Philip's "Invincible Armada" in August had wrecked Spanish seapower and made infinitely harder Spain's task of keeping her overseas lifeline open. The prospects of another long war with England made the fortification of her Caribbean possessions more vital than ever.

Tejeda and Antonelli returned to San Juan in March 1589 with a cadre of skilled artisans, including a dozen stonecutters, 18 masons, a couple of smiths, a cooper, a metal founder, and an overseer. All were on salary and had signed contracts for the duration of the projects. Labor gangs would be picked up at San Juan, Santo Domingo, Cartagena, and Havana. Each place was to supply 150 slaves to be sent by Tejeda to whatever jobs needed them. The expedition also included 320 soldiers, 120 of which would remain in Puerto Rico.

Philip's Council of the Indies had approved three projects for San Juan: a fort on the headland (El Morro); a defensive wall from La Fortaleza across the island to the rocky Atlantic shore; and the sinking of

Philip II succeeded to the Spanish throne in 1556. He vowed to continue his father's policy of trying to keep England from trading in the Caribbean and sent warships and soldiers to enforce it. The freebooter tactics of English sea captains like Sir John Hawkins and Sir Francis Drake tested Philip's mettle throughout his reign. The defeat of the Armada in 1588 forced Philip to strengthen overseas defenses to maintain his New World possessions.

Queen Elizabeth I made her decision to send Drake and Hawkins into the West Indies after she learned that the Nuestra Señora de Begoña, *the 960-ton flagship of the homeward-bound treasure* flota *carrying two million ducats' worth of cargo, was undergoing repairs in San Juan Bay. It was an opportunity too good to pass up, but the long delay in getting the expedition underway gave the Spanish time to prepare a warm reception for Drake's arrival.*

a hulk to block the channel at Boquerón Inlet, the narrow "back door" to the harbor. Antonelli and Tejeda worked closely with Governor Menéndez de Valdés to get the first project started. The other two were tabled for the moment.

The new work on the headland, which Antonelli laid out so as to shelter 3,000 people, was what military engineers called a "hornwork" (two half-bastions connected by a single wall). It would cross the narrow headland and protect the landward side of the tower and Water Battery at the harbor mouth. Tejeda dubbed it "the Citadel." Menéndez de Valdés, who was left to oversee the construction after Tejeda and Antonelli left to begin work on the defenses of Santo Domingo, soon reported with enthusiasm "that the fort, when completed, will be the strongest that His Majesty has in all the Indies. And now the people of the country sleep in security." Perhaps part of his optimism was due to the arrival of another 200 soldiers. In the regular muster of the island military, he now had almost 1,500 fighting men and 80 horses.

Meanwhile, the Crown sent Captain Pedro de Salazar, a veteran of military campaigns in Flanders and Italy, to supervise construction of the new fortification. He reached San Juan on 29 May 1591, with 190 men. Although he and Menéndez de Valdés clashed frequently over money and technical expertise, Salazar was able to get on with the job at hand.

In the new hornwork, one of the half-bastions overlooked the bay and commanded both the harbor and the land approach. It became the dominant position and was named "Austria" in honor of the reigning dynasty of Spain. The half-bastion on the Atlantic side was called "Tejeda." A straight wall (called a curtain) connected these bastions while a narrow moat in front gave added height to the walls and thus strengthened them against assault. From the center of the curtain wall, a gate opened onto a drawbridge that spanned the moat. Both the gate and the bridge were shielded by a small ravelin, embrasured so that its cannon swept the central approach. Behind the hornwork, Salazar leveled off sites for a pair of batteries, one to range out over the sea and the other over the harbor.

Since the hornwork was simply a parapet with no shelter from the weather, Salazar built a guardhouse just inside the gate. He also constructed magazines for munitions—a large one to serve the Austria Bastion and adjacent guns, and another for the Tejeda Bastion area. He mounted the cannon as soon as he could and reassigned troops for the best defensive posture. Then he sent Philip a completion report and a sketch of the new El Morro. The only thing left was to discover how good the new fortification really was.

Sir Francis Drake was the first one to put the defenses to the test. In 1595, Queen Elizabeth gave Drake and John Hawkins joint command of an expedition against Puerto Rico and Panama. The prime target was Panama because it was the great depot for transshipment of New World gold and silver. At first Elizabeth was reluctant to send Drake and Hawkins into the Caribbean again because she believed that Philip II was readying another attempt to invade England. She changed her mind, however, after learning that a homeward-bound treasure galleon had been damaged at sea and was laid up in San Juan Bay for repairs. Its valuable cargo had been ferried ashore and put into La Fortaleza for safekeeping.

Drake and Hawkins left England on 28 August 1595, with 27 ships and more than 2,500 men. The raiders struck first at the Canary Islands, but despite their strength they were not successful. On 28 September Drake set the course for Puerto Rico, which he intended to make a permanent English base, and Canary Islands officials sent a fast dispatch boat to warn San Juan.

Meanwhile, Admiral Pedro Tello de Guzmán had left Spain with five frigates to fetch the treasure stored in La Fortaleza. Near the Canaries he encountered a couple of English stragglers. He captured one, the *Francis*, and chased the other to within sight of Drake's armada. Then, having persuaded the crew of the *Francis* to explain Drake's mission, the admiral made full sail for San Juan. His speedy frigates easily beat Drake across the Atlantic, and his first-hand report gave San Juan's defenders time to prepare for the English fleet's arrival.

Pedro Suarez Coronel, who replaced Menéndez de Valdés as Governor of Puerto Rico, and Captain General Sancho Pardo Osorio, commander of the crippled galleon, were sweating over the treasure in La Fortaleza. They had already planned the best way to meet an attack: build fieldworks at places where the English might try to land and then, when Drake

El Morro at the time of Sir Francis Drake's attack in 1595. Cannon and musket fire from Spanish defenders in the hornwork atop the bluff helped force the English to return to their ships outside the harbor entrance.

1595: Drake's Attack

The Assault on the Frigates

"[A]t ten o'clock at night, when it was quite dark, the enemy commenced an attack on the port with twenty-five boats, each carrying fifty or sixty men well armed, with the view of burning the frigates . . . and they all entered close up to the platform of the Rock [the Water Battery], ranging themselves under the fire of the artillery. . . . Dark as it was, the boats were seen, and instantly the guns from the Rock and from the fort of [Santa] Helena [Elena] began to play as briskly as possible. Most of the boats attacked the Capitana, the Texeda frigate, setting fire to her at the bow, and throwing into her a quantity of fire-pots and shells, while ours succeeded in extinguishing the flames before they had done any damage, the fight being carried on with cannon, musquetry, and stones.

"At the same time they set fire to the [Santa] Ysabel and Magdalena frigates, and to the [Santa] Clara, which was extinguished; but the third time that the Magdalena frigate, of which Domingo de Ynsaurraga was captain, took fire, it was impossible to extinguish the flames, as the ship took fire at the stern and burned furiously; and all that could be done to maintain a footing on board, was done by the . . . captain and the people with him, until the ship was just burnt down and twelve men were killed by the enemy's musquetry, besides as many more burnt. . . . The battle lasted for an hour, the most obstinately contested that was ever seen, and the whole port was illumined by the burning frigate in a manner favourable for the rest, who could thus see to point our artillery and that of the forts, with which, and with the musquetry and the stones thrown from the frigate, they did such effect, that the enemy, after about an hour, . . . retreated with the loss of nine or ten boats and more than four hundred men, besides many more wounded; while on our side, the only loss was that of the frigate and forty men killed or burnt, besides a few wounded by the musquetry. . . ."

—Anonymous, "An Account of What Took Place at San Juan de Puerto Rico . . . on the 23rd of November 1595.:" Translated from original Spanish reports.

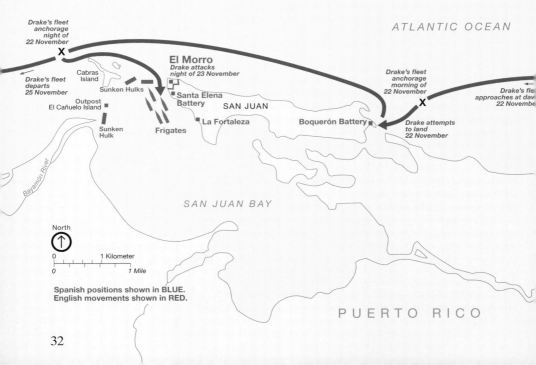

Sir Francis Drake. To the English, he was a hero, a sailor without equal, a man with a nose for gold and a flair for deeds who, in his own phrase, could "singe the beard of the King of Spain!" To the Spanish, he was a devil, and his name was used to frighten children. His death at sea following the defeat at San Juan was celebrated joyfully in Spanish lands. Drake's fame will live forever, but considering the tremendous bulk of Spanish trade, he and others like him were but little fleas on a very large dog. Thanks to Spain's convoy system and her New World defenses, neither Drake nor anyone else had been able to wrest away and hold any important Spanish site, either in the West Indies or on the American continents. This portrait of Drake was painted in 1581 by Nicholas Hilliard, Elizabeth I's court miniaturist.

was sighted, sink the damaged galleon at the harbor entrance to block the channel.

Now the galleon was sunk as planned, along with another vessel. The admiral's five convoy frigates were moored just south of El Morro. There were 1,500 men under arms: 800 from the frigates, the galleon, and the garrison, and the rest local militia. Women, children, and other noncombatants took shelter in the mainland woods.

Coastal lookouts saw the English sails at dawn on 22 November 1595. With his typical boldness, Drake anchored the great fleet within range of Spanish artillery batteries at Boquerón Inlet. But his bravado was costly. One of the 28 cannonballs that the Spanish fired against the English ships crashed into Drake's cabin on the *Defiance*, splintering the stool from under him as he sat at supper. It killed both Sir Nicholas Clifford and a young officer named Brute Browne whom Drake loved as a son. He had already lost Hawkins, who had died of fever before they reached Puerto Rico. But still, Drake would not be distracted from his goal.

By the next morning, the 23rd, Drake had moved his ships beyond El Morro and anchored off Cabras Island. During most of this day his seamen took soundings of the nearby waters, bringing the work boats close to El Cañuelo Island opposite El Morro. But they did not go ashore; the Spanish had a stockade and a small body of troops posted in this area.

That night Drake sent 1,500 of his men to the attack, an even match for Spanish manpower. Armed with "fireworks and small shot" and crowded into shallow-draft pinnaces and other small craft, they forced their way through the port entrance with little regard for either the sunken hulks or El Morro's cannon. In fact, they came so close inshore that Spanish gunners at El Morro could not aim their pieces low enough to hit them.

Swarming onto the frigates, the English set some of them afire. But that proved to be a costly mistake. The blaze lit the battle scene and exposed the English to the Spanish gunners. Musketry from El Morro and cannon fire from Santa Elena Battery and the Spanish frigates cut down the attackers. With muskets and stones, the Spaniards drove Drake's men from the burning frigates into the water.

For an hour the fighting was fierce. Both sides lost

heavily. At the end, as the English drew back to their fleet, four Spanish frigates were still fighting-fit, and the treasure yet lay safe in La Fortaleza. The new defenses, manned by determined men, had withstood even Drake's methodical audacity.

When General Pardo and Admiral Guzmán conveyed the treasure to Spain, San Juan was left with a garrison of 200 men plus some 150 volunteers, instead of the 1,500 troops that had resisted Drake. King Philip knew of Puerto Rico's vulnerability and authorized sizable credits against the Board of Trade at Seville, which was responsible for supervising overseas commerce, and the treasury of New Spain. The first would provide for artillery and other weapons, while the money from Mexico would be used to improve El Morro's defenses.

Unfortunately the credits granted by Philip were just that—credits—and the new governor, Captain Antonio de Mosquera, who came to the island in June 1597, had no funds with which to work on El Morro or anything else. Moreover, his garrison was reduced to 176 men by hunger and disease, and the island population was gaunt from hardship and depleted by death. In San Juan, dysentery had swept away many of the civilians and now only 180 were left. With no labor for the fields, the island was on the verge of famine. Cassava bread and plantains were the only foods to be had. "Even the rich," it was said, "could find only plantains to eat."

But starvation and disease were only part of the picture. The regular subsidy had been held up at Havana, and the men serving in Puerto Rico had not been paid for months. To survive, they sold their souls to the usurers—or became thieves. So grim was the struggle for life among these poor, sick, and starving soldiers that when word came that another great English fleet was crossing the Atlantic in the spring of 1598, Mosquera could do little to make ready. His men, physically weakened, untrained, surly, and hard to handle, had no stomach for building emergency defenses and even less for opposing an enemy force.

This time the enemy leader was another of Queen Elizabeth's favorites, Sir George Clifford, the 39-year-old third earl of Cumberland, noted mathematician and navigator. He had commanded the *Bonaventure* in the defeat of the Spanish Armada a decade earlier, and was a veteran of numerous voyages of exploration and plunder to the West Indies. His fleet of 21 vessels left England early in March 1598, and Cumberland was aboard the *Scourge of Malice*, the most powerful warship of the day.

Again Elizabeth had decided that Puerto Rico must become an English military station where her ships could threaten Spanish lines of communication and seize treasure galleons. San Juan, viewed by the English as "a secure and defensible port," was the target. Cumberland did not intend to repeat Drake's mistake of attacking the well-fortified harbor. On 16 June 1598, he anchored his fleet east of San Juan islet and put his troops ashore west of Condado Point. With a beachhead established and an African slave to show the way, Cumberland led his men toward the only access to San Juan by land, San Antonio Bridge, which spanned the narrow but unfordable Boquerón Inlet. This maneuver, he hoped, would allow him to bypass the main harbor defenses and march directly into town.

Cumberland struck at dawn on the 17th. His 1,700-man army vastly outnumbered the little band of defenders at the bridge. But with the help of the five guns of the Boquerón Battery, which protected both the bridge and the inlet, Spanish pikemen held this critical point through two hours of violent combat. Cumberland himself, attired in heavy armor, sank beneath the bloodied waters and would have drowned had not his men speedily fished him out. The English then withdrew.

Cursing the cannon that had devastated his flank, Cumberland ordered the Boquerón Battery silenced. A warship anchored within range and an hour's cannonading did the job.

Then the English won their second beachhead, this time almost at the foot of the now silent battery. Following some minor skirmishing, the defenders of both bridge and battery, outflanked and hopelessly outnumbered, pulled back into the nearby woods. Nevertheless, the Rev. John Layfield, chaplain to the British fleet, later wrote that the Spanish troops had behaved "gallantly" in their attempt to hold back Cumberland's advancing column.

Before dawn on 18 June the English began the march on San Juan. Moving cautiously in single file for fear of ambush, they entered the town unopposed at daybreak. Only a few Spaniards were there to

1598: Cumberland's Attack

First Contact at San Antonio Bridge "By eight of the clock that Tuesday [morning], being the [16th] of June, his Lordship's regiment, and most . . . of Sir John Berkeley's [the army's second in command] were landed. . . . We began to march as soon as we could be brought into any order. . . . The way we marched was along the sea side, commonly on firm, sometimes on loose sand. . . . [That night, after Cumberland's army had reached its objective,] the Companies were commanded to keep their guard, till his Lordship in person with Sir John Berkeley went . . . to . . . view . . . the place; which they found to be narrow and a long Causeway leading to a Bridge . . . from the one Island to the other [where the town of San Juan was located]. The Bridge they perceived to be pulled up, and on the other bank was . . . a strong Barricade, a little beyond which was a Fort with Ordnance. But how much or what we could not learn, nor by how many men it was held. . . .

"By and by the enemy's Sentinel . . . discovered the approach of our Companies, and . . . [sounded] the Alarm. . . . The assault continued above two hours, during which time the Spaniards were not idle. For though the assailants left no way in the world unattempted, yet no way could they find to enter the Gate. The Causeway, which was the ordinary way of passage, was purposely made so rugged . . . that our men[,] to keep them on their feet, made choice to wade in the water beside it. Here his Lordship was (by the stumbling of him that bore his target [shield]) overthrown, even to the danger of drowning; for his Armour so overburdened him, that the Sergeant Major that . . . was next had much ado . . . to get him from under the water: when he was up, he had received so much Salt water, that it drove him to so great extremity of . . . sickness, that he was forced to lie down . . . upon the Causeway . . . till being somewhat recovered, he was . . . led to a place of . . . more safety and ease. . . . Sir John Berkeley led on his [Cumberland's] Regiment. . . ."

—Reverend John Layfield, chaplain, the earl of Cumberland's army.

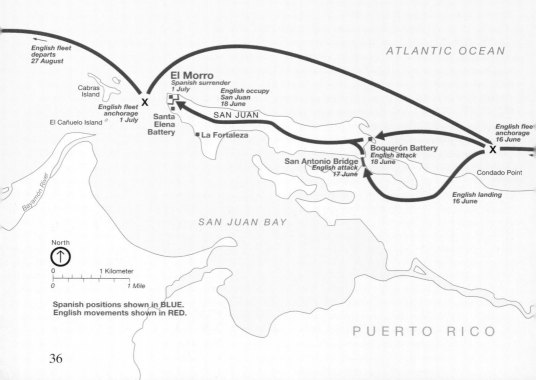

Spanish positions shown in BLUE.
English movements shown in RED.

Sir George Clifford, third earl of Cumberland, was considered "the most prominent leader of Corsairs among the English noblemen." He disliked foreigners, especially Spaniards, who viewed him as a pirate. Before his attack on Puerto Rico, Cumberland had organized 11 privateering expeditions of his own against Spanish possessions and had proposed several others to Queen Elizabeth. He has been called a "wise and clear-sighted courtier, an able military and naval commander, a clever strategist and a brave soldier." His failure to hold San Juan dashed England's hopes of making Puerto Rico a base of operations in the Caribbean. It would be nearly a century before an English military and naval force again threatened the island.

greet them: wounded men who could not be moved, a few women, and old people. The rest had fled inland to the larger island or to El Morro. Cumberland demanded the surrender of El Morro, but Governor Mosquera staunchly refused.

Inside El Morro, however, the situation was already desperate. Mosquera harangued his troops, hoping to rekindle their courage. But nothing could restore the morale of these sullen, undisciplined, and hungry men. The officers had rallied only about 80 men to defend the fort and both water and food were in critically short supply. There was brave talk about a heroic stand, or perhaps even a daring sally to drive the English out of the town. But everyone knew that this was just talk.

Meanwhile, comfortably settled in the town, the earl of Cumberland planned his next move. Deserters informed him of the situation of the tiny garrison at El Morro; they explained how to stop supplies from reaching El Morro from the mainland; they even suggested sites from which siege guns could breach the landward wall of the fort. Cumberland decided not to risk lives on an assault. While his men saw to it that no supplies reached the beleaguered fort, he set up siege batteries on high ground with a good command of El Morro's hornwork. In fact, so well chosen were his two positions that the Spanish guns could not reach them. His fire battered the hornwork, forced the defenders from their parapets, and dismounted their guns. After enduring the shattering bombardment for several days, Mosquera and his war council agreed to negotiate. The white flag was raised over the crumbling walls and the English and Spanish officers met to talk terms.

On 1 July the Spanish governor and his garrison marched out of El Morro with flags flying but armed only with their swords and daggers. All were escorted to La Fortaleza where they would stay until they could be removed from Puerto Rico. Two companies of Englishmen occupied El Morro. Over its walls they unfurled the English standard, the cross of St. George on a white field. Cumberland's armada sailed into the harbor. He was on the verge of making Drake's dream of an English naval base at San Juan come true.

Reverend Layfield was impressed by the captured city. In his eyes, San Juan was an attractive place,

Santa Bárbara is the patron saint of artillerymen. Gunners, like those who manned the San Juan forts, have sought her protection against harm since the introduction of gunpowder into warfare in the 14th century.

warmed by sunshine and caressed by ocean breezes. The houses, he wrote, are built

after the Spanish manner, of two stories height only, but very strongly, and the rooms are goodly and large, with great doors instead of windows for receipt of air. The town in circuit is not so big as Oxford but very much bigger than all Portsmouth within the fortifications, and in sight much fairer.

The Cathedral is not so goodly as any of the Cathedral Churches in England, and yet it is fair and handsome. . . . On the other side [it has] a fair pair of organs. This Church is sacred to Saint John the Baptist, as is all the island. Besides his image there were many others in particular shrines, which the soldiers could not be held from defacing. . . .

There is also a fair Friary standing on the north side of the town. It is built of Brick, in a good large square with a Church and Hall. The Convent was all fled, saving one old Friar who, in the little broken Latin that he had, told me they were of the Dominican Order.

Although Cumberland failed to wring ransom money out of the destitute population, he sent men throughout Puerto Rico in a quick survey of its topography, natural resources, and products, and, of course, to sniff out any signs of gold and to pick up whatever plunder they could lay their hands on. Thanks to the foresight of Governor Mosquera, however, they missed the Chest of Three Keys, in which the subsidy money from Mexico was usually stored. Mosquera, certain of being captured, had removed these funds from La Fortaleza to a secret cache outside the city. Nevertheless, the English collected a considerable amount of booty in ginger, hides, sugar, and slaves, and this despite ambuscades and even open hostility from the island people. Much of the resistance was organized by former Governor Suárez Coronel, who still lived in Puerto Rico.

And now came midsummer's heat and food contamination to wreck Cumberland's plans. Dysentery sickened 400 of his men and killed another 400, leaving him without enough for a garrison. Reluctantly he decided to leave Puerto Rico. The troops pillaged the town and burned much of it; the handsome organ in the cathedral was ripped out and taken away, as were paintings, altar ornaments, and bells from all the churches. The cannon from the forts and even copper kettles out of the kitchens were carted off and stowed aboard the departing vessels.

Philip II learned of the capture of San Juan only three days before his death on 13 September 1598, and he died in the bitter belief that England had at last won her coveted foothold in the Indies. The War Council ordered the recovery of Puerto Rico and readied an expedition of 6,000 men to retake the island. When the news came that Cumberland had withdrawn, the plans were changed: Spain would send 400 soldiers and 46 cannon to replace the garrison and guns carried off by Cumberland, along with supplies and materiel to restore the defenses. And, since the English had also carried off Governor Mosquera, a new governor was named. He was Captain Alonso de Mercado, another Flanders veteran.

When Mercado arrived in San Juan in March 1599, the scars of the siege were still raw. Most of the houses were charred and desolate and the walls of El Morro lay in ruins. The governor's first priority was to strengthen and expand the defenses. The work started almost immediately, and between 1601 and 1609 the hornwork was completely rebuilt upon much stronger foundations, which still support the high and massive walls that stand today.

In 1598, most of the action had been at the east end of San Juan islet where, despite the brave efforts of the Spanish pikemen at the San Antonio Bridge, Cumberland had silenced the Boquerón Battery and outflanked the defenders. During the 1601-09 construction, workmen repaired and expanded these easterly defenses. And to the west, across the channel from El Morro, they built a little wooden fort on El Cañuelo Island and called it *San Juan de la Cruz*. Its firepower would help to control the harbor entrance.

These improvements to the defenses, however, were still inadequate to deter an aggressive enemy, as the events of 1625 were to prove. That year, Puerto Rico found itself confronting the fruits of the growing ambition of another of Spain's ancient enemies—Holland. Dutch merchants, organized in small and large trading companies under franchises and privileges given by the Estates General, had developed a very aggressive policy in which commercial interests and the desire to establish a foothold in the New World combined to undermine Spanish rule in the Americas. In defiance of Spanish regulations against trade with foreigners, the Dutch sold slaves, knives, mirrors, cloth, and flour to Spain's colonies and brought tobacco, sugar, dyewoods, and hides back to their Amsterdam warehouses.

Ordnance

Forts are often described with words like impregnable, unassailable, grim, invulnerable, and redoubtable, terms often derived from fear of a fort's most powerful, long-range weapon: the cannon. A strategically positioned fort with a full complement of weaponry would be a problem for any invader, because the fortress, unlike naval ships, provided a stable platform upon which guns could be mounted and trained on the enemy. Anyone approaching within approximately 500 yards would be in great danger, even though the artillery in those times was not always accurate and aiming was extremely difficult.

Basically all artillery falls into two categories: mortars and guns. Mortars were designed to fire the largest and heaviest projectiles on a curved trajectory. They could shoot over obstacles or fortifications, landing on, and perhaps piercing, the deck of a ship, or hitting a pile of powder kegs or other supplies behind fortified walls, or just wreaking havoc and demoralizing the people. Guns fired their projectiles in a flat trajectory, and their effectiveness in turn depended upon the weight of the shot: the greater the weight of the shot, the greater the damage it would cause.

The first artillery pieces were made of forged iron. The greatest concern was in producing a weapon that could contain the explosive force of the gunpowder, hurl the projectile at the enemy, and not blow up in the faces of the gun crew. Once

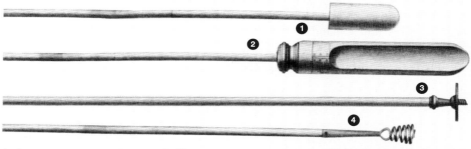

1 Sponge
2 Powder ladle
3 Scraper
4 Worm
5 24-pounder cannon
6 16-pounder cannon
7 12-pounder cannon
8 Grape shot, side view
9 Tongs for handling hot shot
10 Garrison carriage, top view
11 Garrison carriage, side view

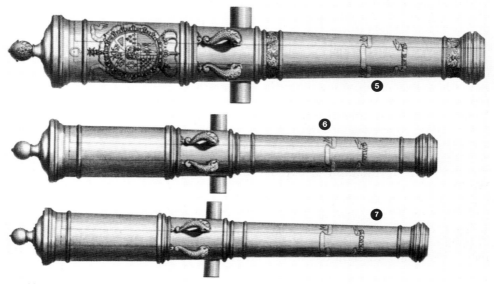

guns could be cast in a single piece in either brass or bronze, great strides were made in the effectiveness of the artillery pieces. By the 18th century bronze seems to have been the metal of choice. The guns and mortars were highly decorated. All bore the coat of arms of the sovereign. Usually the maker was identified in some way; the name might be part of the base ring or shown in a cipher below the sovereign's arms. Garlands of flowers, animals, and mythical creatures sometimes decorated the piece.

All bronze Spanish guns were named—Vindicator, Invincible, Destroyer are a few examples—and the authorities made sure that each gun's whereabouts was always known. This has been invaluable for present-day historians investigating what guns were used where and when.

Guns were classified by the weight of the projectile: a 12-pounder gun shot a 12-pound ball. The kinds of projectiles varied widely: solid shot, canister shot (a container full of bullets), grape shot (a cloth container full of bullets), and bombs or grenades (hollow shot filled with gunpowder) fired from a mortar. Sometimes solid shot was heated until it was red hot. If it landed on a wooden ship, hot shot could set the ship afire. Ordnance enabled a fortification to meet the potential the military engineers had hoped for when they built it. Below are some of the types of artillery and artillery implements used at various times in the forts of Old San Juan.

Tools for Guns

The tools used to operate the ordnance had a variety of functions. The wet sponge swabbed out the cannon to make sure all sparks were extinguished. The ladle dumped the exact amount of powder needed into the chamber. The scraper removed any powder residue. The worm removed unfired bits of cartridge and wadding. The point was to make sure the cannon was clean and free of embers before it was loaded and fired.

These illustrations come from Tomás de Morla's A Treatise on Artillery, *a 4-volume work published in Madrid in 1803.*

How a Siege Works, Circa 1700

1st Parallel

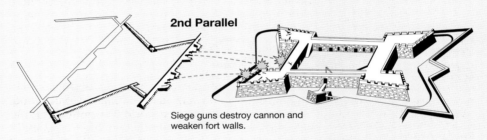

Military engineers, called sappers, construct trenches and raise earthworks to protect attacking forces. Mortar fire destroys cannon and drives defenders to cover. Siege lines prevent supplies from reaching fort.

2nd Parallel

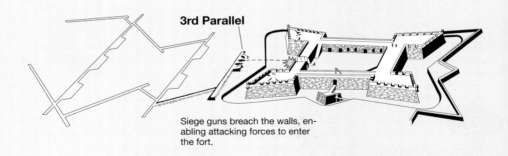

Siege guns destroy cannon and weaken fort walls.

3rd Parallel

Siege guns breach the walls, enabling attacking forces to enter the fort.

A Fort's Defenses

OUTER WORKS				INNER FORT	
Glacis	Covered Way	Moat	Ravelin	Moat	Rampart

Attackers →

Parapet
Scarp
Magazine

The Mechanics of a Siege
Military engineers built forts for several reasons: to protect cities, to protect strong points from falling into enemy hands, to be a visible symbol of governmental authority. If a fort could not be taken by surprise, an attacking party had to take the fort by force. The process of surrounding an enemy's strong point and slowly cutting off all contact with the outside world is known as a siege. Sieges go back to Biblical times, but the principles were formulated by Sébastien le Prestre de Vauban (1633-1707), a French military engineer who served in the armies of Louis XIV. He created a very formal, disciplined science, and his plan was maddeningly simple. First a trench parallel to the fort was dug out of gun range so the attackers could move in supplies and troops. Sappers—crews of trench diggers—then dug zigzag trenches toward the fort; the zigzag pattern made it more difficult for defenders to hit the trenches. Next the sappers dug a second parallel that included some batteries for shelling the fort. Additional zigzag trenches and parallels would be dug until the attackers were in a position to concentrate their fire at one point on the fortification to breach its walls. The fortress would then have no alternative but to surrender or be stormed. Conducting a textbook perfect siege did not always result in success, for the fort's defenders would not have been idle. They would fire cannon at the sappers. Often they dug counter trenches out from the fortress and planted mines to blow up the work of the attackers. And they would send out nighttime raiding parties, too.

But the Dutch were not satisfied with these small successes. In 1624 they seized Bahía, the capital of Brazil, for use as their New World base but could not hold it. A Spanish counterattack drove them out and Dutch reinforcements, led by Boudewijn Hendricksz, the Burgomaster of Edam, came too late. General Hendricksz then turned to Drake's old dream: the capture of Puerto Rico.

The Spanish were warned of a possible Dutch strike, but the appearance of Hendricksz's 17-ship armada still caught them by surprise. As the call to arms sounded throughout the city, Governor Juan de Haro, a soldier well-experienced on the battlefronts of the Netherlands as well as the Caribbean, took the routine steps of sending out patrols, scouting the environs, collecting supplies, and recruiting militiamen. Noncombatants were sent to the countryside. And with much fanfare the troops were marched from the town square to their posts in full view of the enemy, so that he—whoever he was—might be impressed with their numbers.

Governor Haro, however, was not optimistic. He was short on powder, fuses, weapons, and food. "There is no fortress in the world," grumbled this old veteran, "which can be defended without victuals."

Before a strong favoring breeze, Hendricksz's fleet formed into battle line at one o'clock in the afternoon of 25 September and sailed boldly for the port entrance. They saw El Morro on the headland, but they gambled that its guns, even if manned by expert artillerymen, could not keep them out. Hendricksz's flagship led the line. To confuse the Spaniards, the Dutch ships flew ensigns of many nations. El Morro's gunners were not fooled, however, and they opened fire on Hendricksz as soon as he came within range. Hendricksz returned the fire and his ship was soon inside the harbor with the rest of the fleet. Haro later confessed that his gunners were "tailors, cobblers and other artisans," few in numbers and without battle experience, and they simply could not cope—especially since their cannon, when fired, tended to split their ancient carriages.

The other side claimed that Haro's tailors and cobblers put up such a fierce defense that the harbor was won only by a most heroic display of Dutch fearlessness! In any case, Hendricksz's personal courage excited even his cautious officers and brought the

1625: The Dutch Attack

El Morro Besieged "That night [Thursday, 25 September, in anticipation of a lengthy siege] there was placed inside the Fortress 120 loads [756 bushels] of cassava bread, 46 *fanegas* [153 bushels] of corn, 130 small bottles of olive oil, 10 barrels of biscuits, 300 Canary Island cheeses, 1 hogshead [126 gallons] of flour, 30 jugs of wine, 200 birds, 150 boxes of quince meat, 50 beeves [which] ... had been brought to El Morro by the town councilman Francisco Daza who had been entrusted with them, along with 20 horses. With these provisions and succor, that same night before the enemy set foot on dry land and blocked the passages to the countryside, the Governor [Juan de Haro] named as quartermasters Captain Alonzo de Figueroa, Warden of the Castle [El Morro], Francisco Daza and Diego Montañes, City Councilmen, Captain Pedro de Villate, Juan de Lugo Sotomayor and Don Juan Ponce de León, giving to each a commission to seize canoes, boats, and whatever kind of vessel, and [use] them to aid the fortress, with whatever supply of meat and cassava bread and corn that they might find.

"Tuesday the 30th [at about 9 o'clock in the morning when the Dutch were ready to besiege El Morro], there came an enemy drummer with a white flag and a letter [demanding that Governor Haro] 'deliver over into our hands the Castle with its people' [or suffer the consequences. The governor responded:] 'If Your Grace might wish or want to try something, it should be to deliver over to me your vessels that are anchored in this port, of which I will give you back one as it might be needful so that you can go away; this is the order that I have from my King and Lord, and no other.'

"And the enemy having seen the answer to his letter caused to be fired ... more than 150 rounds of artillery. God was served that none of ours was injured."

—*Diego de Larrasa, Lieutenant Auditor General, November 1625.*

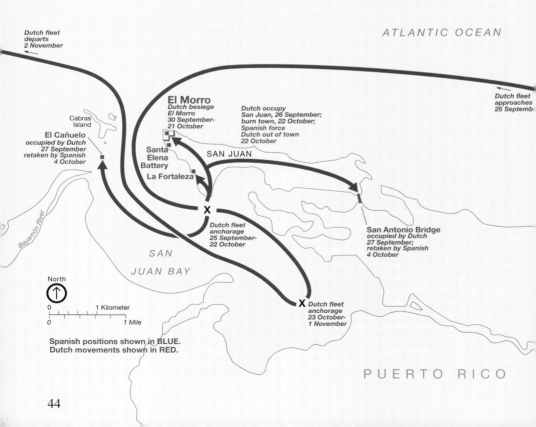

44

Dutchmen into San Juan Bay with the loss of only four men and a few wounded aboard the flagship.

By the time the fleet had passed El Morro, the afternoon was well along; and since there was a shoal between the ships and the landing area south of the town, the methodical Dutchmen would not risk running aground in their haste to get ashore. Landing was postponed until the next morning.

That afternoon and all through the night the townspeople scurried about, packing their best belongings, and, with a prayer heavenward, fled inland over San Antonio Bridge and out of the enemy's reach. A detachment also hurried to the Puerto Rican mainland to call men to arms and bring in food supplies for the garrison. At sunset the governor marched the garrison into El Morro. There were only 330 men and he could not afford to lose any. That night the fort was stocked with all available provisions: cassava, maize, olive oil, hardtack, cheese, flour, wine, 50 head of cattle, and some chickens.

At dawn on the 26th, the Dutch fleet opened fire on the town as small boats filled with assault troops headed shoreward. General Hendricksz was the first to leap ashore. Some 800 men were soon with him, and he marched them in close order into San Juan as far as the market place at the center of town. There was no resistance: all the people were gone.

Hendricksz chose La Fortaleza for his headquarters and raised the banner of the Prince of Orange over its ramparts. His soldiers marched from the market place to the cathedral, where they toppled the statues, trampled the furnishings, and carried off anything they thought of value. Then they lodged themselves in the city's best houses. Hendricksz, incidentally, imposed heavy penalties on drunkenness; many of the wines liberated from Spanish larders were poured out by his orders.

It took three days for General Hendricksz to prepare to besiege El Morro. He blocked the eastern approach to the fortress by placing troops at the San Antonio Bridge and on El Cañuelo Island to the west. Since the fort on El Cañuelo controlled the mouth of the Bayamon River, El Morro's main line of communication with the island's interior, this was a devastating loss to the Spanish. It meant that El Morro was cut off from inland support and supplies.

With El Cañuelo and San Antonio Bridge in hand,

Hendricksz began the siege trenches for his sappers (military engineers) to use in approaching the hornwork. On the 30th Hendricksz delivered an ultimatum to the governor: *Surrender the fort or I will take it and put all to the sword, even the women and children.* Haro refused; the Dutch approach trenches (called "saps" by military engineers) inched closer. The heavy Dutch guns were manhandled into place, and the siege began. For 21 days the duel continued. Dutch cannon fired more than 4,000 balls, causing heavy damage to El Morro's massive hornwork. The tailors, cobblers, and others who manned the Spanish guns learned from a Dutch prisoner that their cannonading was also very effective and that many of the Dutch had been killed or wounded.

While Hendricksz applied the pressure at El Morro, Haro's people released it elsewhere. At San Antonio Bridge, Puerto Rican militiamen harassed the Dutch garrison and finally drove it out. Captain Andrés Botello and his partisans regained control of the Bayamón River, then stormed El Cañuelo, captured the Dutch defenders in a hot two-hour fight, and burned the fort. Again the way was open to reach El Morro with supplies.

Still Hendricksz pushed the siege trenches relentlessly toward the hornwork. They reached the glacis before the wall, and it would be only a matter of time before Dutch cannon were close enough to breach the wall.

Perhaps the successes of the Puerto Rican militia inspired the governor and his garrison to greater effort. One of his gunners blasted one target after another with his cannon. Companies from the garrison made several sallies from the fort, and one under Captain Juan de Amézquita drove the besiegers out of their trenches.

On 21 October, in a last effort to win El Morro, Hendricksz sent the governor another ultimatum: *Surrender, or we burn the town.* Again Haro refused: "We have enough courage and wood and stone to build again."

That day the invaders went through the town house by house. When they were through, nothing of value was left. Everything worth taking went aboard the ships: sugar, ginger, tobacco, skins, clocks, brassware, coins, guns, provisions, slaves; and from the churches a crucifix and altar ornaments, and, of course, the bronze bells. On the 22nd they torched the buildings, public and private alike.

As the smoke from the city billowed into the sky, Haro loosed a desperate two-pronged attack on the Dutch troops. Captain Amézquita headed the force from El Morro; Captain Botello led his *puertorriqueños* from San Antonio Bridge. Together they routed the Dutch from their positions and killed or captured many. The rest withdrew to the ships. Then, from the bay shore, Spanish gunners (no longer mere tailors and cobblers!) fired at the enemy fleet. They claimed 13 hits on the flagship alone, and heavy damage to other vessels, before the Hollanders moved out of range. The rest of the month Hendricksz kept his ships at a distance in the inner harbor, making repairs and waiting for favorable sailing weather.

Haro had planned to bottle up the enemy by stretching a heavy chain of logs across the port entrance, but the chain was not yet finished when the Dutch fleet raised anchor and headed for sea on 2 November. Hendricksz and his captains swept past El Morro and out into the Atlantic. But they did not go scot-free. Haro's soldiers had realigned the cannon at El Morro, putting them on platforms facing the passage at four different levels, and the Spanish gunners left many a scar on the Dutch ships as they passed by. And scarred, too, was Haro. Thanks to an enemy hit on an artillery rammer, he plucked 24 wooden splinters out of his skin.

Hendricksz left San Juan in ashes. The Dutch arsonists had done their work well. Ninety-six buildings were burned. Of La Fortaleza, only the master walls remained. Civil and church archives were irretrievably lost. The handsome house of the bishop, with its priceless library, was gone and most of the Dominican convent was blackened debris.

The Dutch attack proved that an aggressive enemy could bypass El Morro and still take the town and its outlying defenses. True, determined defenders could win them back—but at a high cost. The lessons learned in 1625 strongly affected future development in Puerto Rico. It was clear that there had to be even more defenses—and soldiers to man them—if San Juan and Puerto Rico were to be saved for Spain.

The Spanish had always been aware of the possibility of invasion by rival European powers, and the danger level rose as the French, the Dutch, the

Danes, and the English planted colonies among the Caribbean isles. The nations refused to acknowledge Spain's monopoly in the Americas. They considered their colonies to be legitimate and they were prepared to exploit the resources of the lands they occupied and look for trade wherever they could. Many were not opposed to practicing a little piracy as well.

Every enemy base, whether under a national flag or the black flag of piracy, was a thorn in Spain's side. To some degree, each one flanked not only the treasure route but also Puerto Rico, Hispaniola, and Cuba, as well as the mainland of South America. "Intrusive" French and English settlements in North America also disturbed Spain, especially as they inched southward toward the Spanish Florida peninsula and the strategic Bahama Channel used by the fleets on their return to Spain.

Though possessing strength far beyond her size, in truth Spain was dangerously overextended, short on manpower and shipping, and unable to stop intruders from preempting lands in the vast Americas. And this lack of manpower opened her colonies to foreign traders—the *contrabandistas*. If Spanish vessels could not meet colonial needs, who could turn the smuggler away? Especially when his goods were needed and the prices were right! As the years went on, the San Juan military was ordered to put a stop to contraband trade, but this was an impossible assignment as long as the island people greeted the *contrabandistas* with open arms.

Contrary to official fears, contacts with foreigners did not "corrupt" the *puertorriqueños*. They showed fierce loyalty to Spain in the 1625 attack, and from 1635 to 1655 they manned no less than six expeditions against English, French, and Dutch colonies, all the way from Tortola, one of the Virgin Islands, to Jamaica. By the last quarter of the 17th century, Spaniards had begun to practice some of the lessons taught them by the buccaneers and smugglers. Puerto Rico had become a leading center of Spanish privateering operations against foreign shipping in the Caribbean and brought in many prizes.

The old world meets the new. An 18th-century bronze cannon bearing the coat of arms of the Spanish sovereign thrusts its muzzle through one of Fort San Cristóbal's ancient embrasures towards the modern city of San Juan.

Part 3

A "Defense of the First Order"

This coat of arms, which hangs above the entryway to El Morro, is that of Charles III, who was king of Spain when the San Juan forts were being completed in the last half of the 18th century. The Spanish government presented it to San Juan National Historic Site in 1964.

Previous pages: *Dutch engraving showing San Juan about 1625. La Fortaleza appears in the center of the illustration, El Morro in the distance at left.*

After the 1625 attack there was a burst of construction at San Juan, stimulated by changes in the political map as other European nations planted their flags in the Indies. The Spanish crown reaffirmed Puerto Rico's strategic significance in 1645: "It is," said Philip IV, "the front and vanguard of all my West Indies, and consequently the most important of them all—and the most coveted by my enemies." It must be protected.

El Morro had to be repaired and strengthened. The work went slowly, however, because rebuilding La Fortaleza, the people's houses, and other structures had to be done at the same time. As usual there was not enough money or manpower. By the 1650s, however, El Morro took on the look it would have for the next hundred years. On the land side loomed the hornwork, now restored with moat, drawbridge, and ravelin. Encircling the headland and following its contours was a barrier wall. Inside the enclosure were the several batteries and a cluster of small structures used variously for quarters, magazines, kitchen, hospital, chapel, cisterns, and such.

The new city, built upon the ashes and rubble of the old, numbered about 250 houses, many constructed of masonry instead of wood. To protect the new San Juan, the Spanish planned to surround the town with a great wall (except where natural battlements already existed on the Atlantic side). They would also build a fort to prevent another Cumberland from attacking by land from the east. This fort, named San Cristóbal for the rocky hill on which it stood and the precursor of the great *castillo* that would one day occupy the site, was situated at the northeast end of the city wall, on the location recommended long ago by Diego Menéndez de Valdés.

Construction of the wall got under way on 26 July 1634. Crown money started it, but the Town Council had to raise most of the funds and did so by levying high taxes on ginger, tobacco, and wine—an action that, according to angry citizens, almost bankrupted the townsfolk. The citizenry also had to furnish labor. This was a slim crew at times, for recurring epidemics decimated both the city and island populations.

By 1650, masonry walls closed in the town on the east, south, and west and included San Cristóbal overlooking the eastern approaches. At this time, San Cristóbal was hardly more than a semicircular platform for cannon nestled between two salients of the town wall, and barely hinted at the mighty fortification it would one day become. El Espigón, a sea-level salient below San Cristóbal now renowned in Puerto Rican legend as *la Garita del Diablo* (the "Devil's Sentry Box"), may be a remnant of this or an even earlier construction.

In the 1660s there was no wall along the rocky Atlantic coast, and to protect this sector the defenders built a small fort between El Morro and San Cristóbal. They called it *La Perla* (The Pearl), borrowing the name from the nearby sandstone quarry where the old cemetery is now. (The fort is gone, but the area below the city walls where it once stood is a *barrio* that still carries its name.)

At the far eastern end of the islet, the small fortification at Boquerón Inlet was rebuilt in 1646. Later it became San Gerónimo. In the early 1660s, on the tiny island of El Cañuelo opposite El Morro, Governor Pérez de Guzmán erected a small, permanent fort. It was a 50-foot-square masonry battery built on rocks bulkheaded by timbers. About 1681 the fort at San Antonio Bridge was reconstructed, and the bridge itself reinforced with a new drawbridge. San Antonio and San Gerónimo were isolated by wet moats as protection against land attack.

By the beginning of the 18th century, San Juan was one of the better fortified places in the Indies. This was fortunate, because the year 1700 brought a new alignment of the European powers as the grandson of France's King Louis XIV came to the Spanish throne as Philip V. The new Franco-Spanish alliance drew Spain into a long series of wars against Great Britain. Much of the fighting was at sea, and Puerto Rico, geographically and historically in the path of clashing imperial interests, was open to attack. Alarms, rumors of raids, and the presence of enemy fleets in nearby waters kept the people of Puerto Rico permanently on alert. There was little time and few resources for major defensive projects during the first half of the 1700s, but the fortifications were maintained and some works even strengthened.

Through these trying years the San Juan garrison remained small, even though defense responsibilities grew. Indeed, official neglect of the men-at-arms had permitted many of them to develop more interest in trades, farming, and stock-raising than in soldiering.

El Morro

The full name of this massive fort is *Castillo de San Felipe del Morro*, meaning "Fort St. Philip of the Headland." Named in honor of King Philip II of Spain, it was designed by Juan Batista Antonelli and built by 18th-century engineers on the site of earlier fortifications. It covers more than 5 acres. Storerooms, gunrooms, officers' and soldiers' quarters, a chapel, and a prison surround a large courtyard, or assembly plaza, and huge cisterns lie beneath. Ramps, tunnels, and stairways afford access to the different parts of the fort.

Harbor defense was El Morro's main mission, and El Morro formed the nucleus of the San Juan fortifications. In its heyday, its batteries of cannon on four different levels were deadly deterrents to any enemy warship that dared to enter San Juan Bay. At sea level beside the entrance channel, guns in the water battery could break the waterline planking of passing vessels or fire "hot shot," cannonballs heated until they glowed cherry-red, into a ship's hull, starting fires.

The cannon at El Morro's second level (the casemated guns) faced directly seaward, and gunners here would try for hull and deck damage; with any luck they might cut a mast or two. Third-level gunners (in the Santa Bárbara Battery) aimed at sails and rigging, and cut them to pieces with canister and chain or bar shot. Fourth-level guns (the Carmen Battery) supported the Santa Bárbara Battery and included in their field of fire the entire western and even part of the eastern sector of San Juan.

Because El Morro was vulnerable from the rear, a fifth tier of guns pointed landward, mounted on a barrier wall called the "hornwork" because its plan resembled the outreaching horns of a bull—two half-bastions (the horns) joined by a curtain (straight) wall. In front of it was a barrier moat. Beyond that was cleared land called a glacis, smoothed and sloped so there was no shelter for attacking troops from the gunners and musketeers who lined the fort walls.

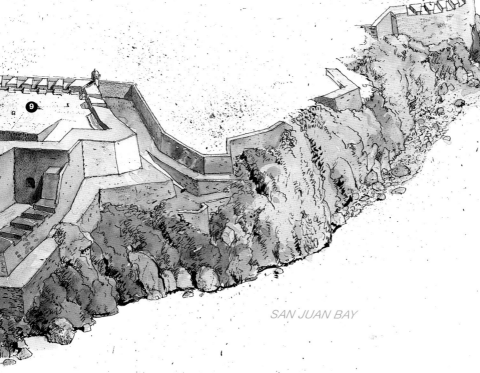

SAN JUAN BAY

1 Water Battery
2 Santa Bárbara Battery (enclosing original 1540 tower)
3 Casemate Guns
4 Carmen Battery
5 Ochoa Half-Bastion
6 Hornwork
7 Drawbridge
8 Moat
9 Austria Half-Bastion
10 Glacis
11 City Wall

Aspects of a Soldier's Day

At each dawn, the loud clatter of drums and the shrill rising notes of fifes playing "*La Diana*," the Spanish "reveille," signaled the troops of Old San Juan that the working day had begun. For the soldiers at El Morro and San Cristóbal, it was the start of another cycle of work details, drills, and ceremonies—a daily routine whose predictability brought a measure of comfort and stability to their lives.

Fife and drum punctuated the soldiers' day, calling them to the *plazas* to form squads and companies to render honor to

Dishes for the soldiers' table.

The chapel at El Morro.

the flag, witness punishments, or attend mass in the chapels. They spent hours learning how to load and fire their flintlock muskets and smoothbore cannons "by the numbers," a sequential drill procedure that made it easier to master these blackpowder weapons, promoted better teamwork and helped to build confidence and discipline in the ranks.

Although the weapons technology and drill techniques used by the San Juan garrison followed the latest developments in Europe, the food the soldiers ate was far different from that of their European counterparts. The most common meal was a one-pot stew made by slowly boiling beef or pork for hours with chunks of pumpkin, corn, okra, green bananas, plantains, and various tropical tubers such as *yautía, malangas*, and sweet potatoes, seasoned with garlic, coriander, onions, sweet peppers, and annatto. This "*olla podrida*" (rotten pot), which originated in 16th-century Spain, has evolved into the still-loved *sancocho* of Puerto Rico today.

When the working day ended, usually in late afternoon, off-duty soldiers could light up *cigarros* and amuse themselves with card games or dominoes until the rhythmic sounds of "*La Retreta*" sent them to their barracks. After the forts were locked and the lookouts posted at the *garitas* along the city walls, citizens and soldiery considered Old San Juan secured for the night, safe behind its massive walls and stout gates.

Fortunately, the island's militia stepped into the breach left by the regular soldiers. They beat off raids on several coastal towns and three times helped the garrison stop British attempts to settle the neighboring island of Vieques.

Since appeals to Spain for more protection were fruitless, the Spanish governors of the American provinces licensed local sea captains as privateers or *guarda costas* to protect the coastal towns and to cope with alien settlement and shipping that Spain had decreed illegal. These privateer captains brought dozens of foreign prizes into Spanish ports to be condemned and sold. Puerto Rico, thanks to its location, became a major privateering base. Unhappy British officials in the Caribbean several times proposed to wipe out this "nest of pirates" and then use the island for the protection of British commerce. So aggressive and effective were the *guarda costa* that London merchants howled loudly enough to pressure Parliament into a declaration of war against Spain in 1739. This conflict was briefly called the War of Jenkins' Ear, in honor of British shipmaster Robert Jenkins, who claimed to have lost an ear in an encounter with Spaniards off the coast of Florida; but it quickly merged into the War of the Austrian Succession (1740-48).

Although the activity of the Puerto Rican *guarda costa* made retaliation probable, there were no large-scale operations against the island. Most of the fighting took place in Cuba and on the Spanish Main. Later, during the Seven Years' War (1756-63), the major fighting took place in North America. Britain won Canada from France and Florida from Spain and increased its holdings in the Caribbean. Both France and Spain, however, hoped to regain their losses.

Spain's King Charles III, a hardworking monarch who came to the throne in 1759, made innovations that reached deeply into administration, economics, and the military throughout his domains. He realized that the overseas empire had to carry its share of the defense load before Spain could reclaim her losses, for a weakly defended Spanish America made Spain also vulnerable. As a first line of defense, Charles decided that the defenses of Caribbean ports had to be strengthened and modernized. Since the British captures of Havana and Manila in 1762 had shown that relying only upon fixed fortifications was folly, so the new idea was to organize colonial armies to bolster naval and harbor defenses. The record of the militia in Puerto Rico and elsewhere must have influenced this decision. The core of the new armies would be either units rotated from Spain or professional troops raised and stationed permanently in the colonies. To these professional soldiers would be attached an enlarged and well-trained militia.

To carry out such radical changes in the face of entrenched officialdom, Charles III relied on intelligent and energetic—and sometimes ruthless—executives such as Alexander O'Reilly, a Dublin-born Irishman. He had joined the Spanish army as a lad and fought the Austrians in Italy. Afterward he served in both Austria and France, then returned to Spain for the war with Portugal. In 1763, O'Reilly had come to the Caribbean to reform defenses and choose a port to be fortified as a base for the Caribbean fleet. His title was Inspector-General of Cuba, and his first chore was to rebuild and modernize the Havana defenses, dismantled by the British during the Seven Years' War. This he did, along with reorganizing Cuba's armed forces.

In April 1765, Field Marshal O'Reilly came to San Juan. When he stepped ashore, one of the first to greet him was the chief engineer of the presidio who, like O'Reilly, was an Irishman. His name was Colonel Thomas O'Daly and, like O'Reilly, had enlisted as a boy in the Spanish army. O'Daly began as a lowly apprentice in the Corps of Engineers and rose to field engineer. In 1765, at 37 years of age, he was appointed Chief of Engineers at San Juan.

Alexander O'Reilly remained in Puerto Rico only 45 days, but during that short stay he, together with the governor, O'Daly, and O'Daly's aides, gathered information for remarkably perceptive reports dealing with economic, social, and defense problems, and even set forth the reforms he felt were needed to correct the island's economy.

He found graft and corruption among garrison officers and lax discipline among the soldiers, who were more interested in scratching out a living for their large families than fighting for their king. They had no military quarters but lived on their own in poverty. Many were sick or invalided.

O'Reilly's remedies were drastic. He discharged those unfit for duty and replaced married soldiers

Since wheat, barley, and rye will not grow in tropical climates, and the closest source of bread flour was several months' sea voyage distant, the European custom of eating bread at every meal was not followed by troops of the San Juan garrison. They and other Spaniards in the 16th-century Caribbean quickly adopted cassava, a dry cracker-like bread Taíno Indians made from the grated roots of the yucca or manioc plant (above). Cassava was better than wheat bread hardtack for military rations or naval stores, since it was almost impervious to tropical dampness, lasting several months without mold. Cassava bread is still available today in San Juan markets.

Spaniards in the Caribbean were the first to produce and export sugar, derived from the sugar cane plant (above). The growing and seemingly insatiable demand for sugar in Europe made it the "green gold" of the West Indies in the 17th and 18th centuries, and it was one of the primary reasons the Dutch, the Danes, the French, and the English were so eager to acquire and hold Caribbean real estate. It was also a major factor in the introduction of slavery into the area. Sugar plantations were lucrative and brutal businesses that generated a huge and constant demand for workers. The Spaniards saw the African slave as the logical solution to meet that demand and other labor shortages in the New World.

with single men. He mustered the best soldiers into a new battalion that replaced the old garrison and was given the same status as a regular unit from Spain. In 14 key towns of Puerto Rico, O'Reilly's plan organized infantry and cavalry militia totaling more than 2,000 men. Their mission was to guard the island coasts and to reinforce the San Juan garrison in the event of an invasion.

The two Irishmen also prepared a project for strengthening the San Juan fortifications. O'Reilly reported to the Crown that Puerto Rico was the crossroads of America, not only for Spain but for its enemies as well. Its windward position made the island the best place for a base from which to aid (or invade) Spanish America. Charles III was so impressed with O'Reilly's analysis that on 26 September 1765, he declared San Juan a "Defense of the First Order." The next day he approved the O'Reilly-O'Daly project to make it so.

The Royal decree spelled out San Juan's functions: it must, of course, protect Puerto Rico and would, as in the past, be a port of entry and place of acclimatization for people and plants coming from Spain. It must also be a base depot and naval station to support and help secure Spain's commerce. It must be the bastion of the Antilles and an outpost of the Gulf of Mexico. Challenging the imagination of his planners, Charles III added that San Juan as a defense of the first order must stimulate progress in industry, agriculture, and the arts, which are "the basis of a nation's real wealth."

O'Reilly's 45-day stopover in Puerto Rico stimulated a long-lasting building and economic boom in San Juan. Its strategic importance now fully recognized, this island city became one of the strongest—if not *the* strongest—fortified cities in the Americas. The pair of Irishmen, along with the king's money, brought a golden age of architecture to San Juan. In combination with spectacular fortifications rose other structures—military, ecclesiastical, and secular—designed by talented men whose signatures verify the meticulously drawn construction plans.

O'Reilly's critique of the old fortifications had pointed out that the only real defense for the town was the wall built after the 1625 attack. Once it was breached, the town was lost. So for 25 years and more the engineers and countless laborers under the direction of Thomas O'Daly and his successors, worked to give San Juan a defense-in-depth system that could withstand any invader. Their work centered on coastal and harbor batteries emplaced in El Morro, San Cristóbal, certain salients of the city wall, San Gerónimo, and El Cañuelo; and land defenses that included the works in the Boquerón sector, the line east of San Cristóbal, and completion of the city wall along the Atlantic from San Cristóbal to El Morro.

Despite unprecedented amounts of men, money, and materials, it was not until the early 1790s that the fortifications were completed essentially as they remain today. San Cristóbal and El Morro alone took more than 20 years to build. O'Daly never saw his project finished. He died in 1781 at the age of 53 after a lingering illness and was buried in the Cathedral of San Juan. His friend and collaborator since 1766, Juan Francisco Mestre, completed the work.

San Juan's defensive makeover began on 1 January 1766 at San Cristóbal, the most exposed sector of the old fortifications. In contrast to El Morro, where the builders could take advantage of the steep terrain to erect the vertical defense needed at the port entrance, the San Cristóbal topography had to be greatly modified before the defense-in-depth principle could be achieved. The terrain had to be adapted to the fortification—and this in a day before dynamite and bulldozers. And as the work moved along, the people of San Juan must have marveled as the walls loomed ever higher.

East of San Cristóbal, workers cleared and graded the ground so fort guns could reach everywhere. Hollows and hedgerows, ditches and trees disappeared in the leveling operation. On the bay side south of the fort, some of the swamps were filled in while others were left as natural barriers.

The labor was staggering. At San Cristóbal, more than 400 men a day were employed at the peak of construction. By the king's order the labor gang included not only local day laborers but convicts (thieves, killers, rapists, and the like from the jails of Cuba and Mexico) as well as soldiers serving at that time in San Juan. Slaves labored principally in stone quarries opened near the construction sites. Hundreds of civilians were employed in supplying building materials such as timber, lime, brick, sand, and water to the different sites.

San Cristóbal

Here, as at El Morro, the engineers founded their defense on a natural promontory. *Castillo de San Cristóbal*, or "Fort St. Christopher," is the largest of the San Juan forts. It rises some 150 feet above sea level on the northeast edge of the old city about half a mile from El Morro. While El Morro's chief mission was harbor defense, San Cristobal's was defense against attack by land. The need for protecting the city's landward approaches was demonstrated by the earl of Cumberland's attack on San Juan in 1598 and by the Dutch attack of 1625.

San Cristóbal was begun in 1634 as a small triangular-shaped redoubt and modified over the years. By the time it was finally completed in 1783, San Cristóbal had developed into a complex of fortifications covering about 27 acres of land. Its shoreline location made the construction of additional coast-defense batteries necessary on the seaward side to defend the fort from naval attacks. Though some of its outworks have been lost due to the expansion of the city, it remains a spectacular example of the defense-in-depth principle.

The great improvement in San Juan's defenses in the 18th century was due largely to the

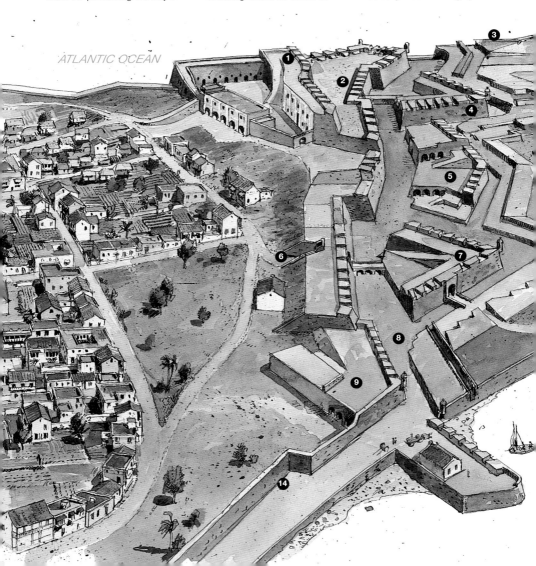

efforts of two men: Spain's King Charles III *(near right)*, the "Enlightened Monarch" whose reforms during his reign made San Juan a "Defense of the First Order," and Charles's representative, the Irish-born Field Marshal Alexander O'Reilly *(far right)*, who considered Puerto Rico the crossroads of Spanish America and the best place for a base from which to aid (or invade) Spain's New World possessions.

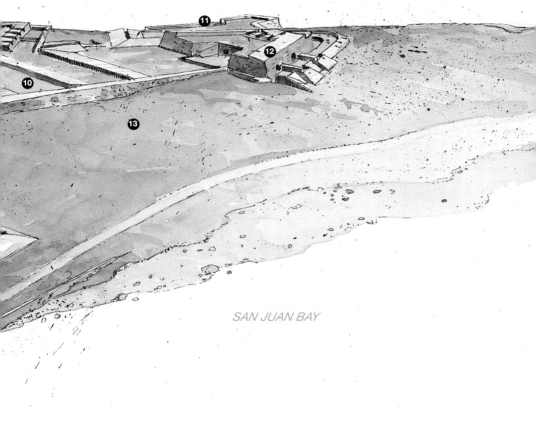

SAN JUAN BAY

1	Cavalier	8	Moat
2	Main Battery	9	Santiago Bastion
3	Santa Teresa Battery	10	Plaza de Armas
4	San Carlos Ravelin	11	La Princesa Battery
5	Trinidad Countergard	12	El Abanico
6	Santiago Gate	13	Glacis
7	Santiago Ravelin	14	City Wall

Of Cisterns and Bulwarks

Engineers Thomas O'Daly and Juan Mestre confronted many difficult problems during the building of San Cristóbal and in making improvements to El Morro in the mid-18th century. One was how to provide water for thousands of soldiers and civilians during a long siege in tropical heat when there is no nearby river, lake, pond, or other source of fresh water. To solve this problem they built cisterns under the main plazas of each fortress, where, thanks to gutters and pipes and the laws of gravity, rain was collected from the roofs and floors and stored until needed. El Morro's three cisterns could hold a total of 216,000 gallons of water, while San Cristóbal's five cisterns, one of which is shown below, could accommodate 716,000 gallons. Together these seven cisterns could store nearly a million gallons of water, about half of San Juan's annual rainfall.

Another problem O'Daly and Mestre faced was how to build a fort wall that would absorb the impact of a 24-pound cannonball traveling at more than 1,000 feet per second. (That's enough force to penetrate four feet of solid oak boards nailed together.) The answer was to make the walls 18 to 40 feet thick, which they did at both El Morro and San Cristóbal. To conserve time and expense, O'Daly and Mestre made the main walls of the fortresses a sandwich of hard-soft-hard materials, as depicted in the illustration at right. The exterior or outer walls (those that would be facing an enemy) were made of custom-cut blocks of sandstone and limestone, of modest thickness, tapering from base to crest. The large space between the outside laid stone wall and the thinner interior laid stone wall was filled with *mampostería*, or "rubble-work." Chunks of stone trimmed off the main building blocks, broken bricks, and fragments of pottery were mixed with a wet

1 Entry ramp
2 Plaza
3 Well
4 Officer's quarters
5 Cistern
6 Chapel
7 Soldiers's barracks

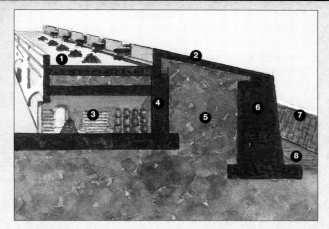

mortar of sand, lime, and water and poured into the cavities left between inner and outer walls, hardening into an economical and effective filler. Should a cannon ball pierce the hard outer laid stone wall, the softer *mampostería* would absorb whatever energy it had left and prevent it from penetrating to the interior of the fortress.

These and many other construction features of El Morro and San Cristóbal continue to delight those visitors curious enough to take a closer look, read the educational displays, and bring their questions to the ranger staff at the forts.

1 Terreplein
2 Parapet
3 Storage or quarters
4 Inner wall (masonry)
5 *Mampostería* (fill)
6 Outer wall (masonry)
7 Counterscarp
8 Moat

There were also great changes at El Morro. Although improvements following the Dutch attack of 1625 had made this a strong multi-level defense, the work of O'Daly and Mestre now converted it into a superfort. From the ancient Water Battery at the foot of the headland to the massive hornwork above, El Morro was modernized and expanded. The old design, with buildings placed here and there almost haphazardly, gave way to a more logical and unified arrangement (see pages 54-55).

The core of the new work was its main battery, called Santa Bárbara. It continues to dominate the harbor entrance like the prow of a mighty battleship. This battery evolved in two major steps from the little cliff tower of the 1500s. In the 17th century the original tower was enclosed in the masonry of a U-shaped battery. A century later this battery was enlarged into Santa Bárbara. Firepower on this level grew from the three cannon of the original tower to 11 in the 1600s. In the 1780s, the number of cannon increased to 37 or more, not including the casemate guns on the Atlantic side at a slightly lower level.

Behind the Santa Bárbara Battery a huge wall swept upward almost vertically from bedrock to the topmost level of the hornwork. This tremendous scarp not only shielded the rear of the hornwork, the living quarters, magazines, and cisterns but also contained casemates for cannon at a higher level than Santa Bárbara. This great wall, with its grim and impregnable aspect, discouraged attack and gave unity to a beautiful example of engineering art.

The hornwork was changed. Redesigning the flanks by filling in the orillons or "ears" preferred by earlier engineers made more room in the bastions. Old parapets were replaced by thicker ones pierced by 26 embrasures for cannon—almost twice as many as before. Bombproof rooms built against the hornwork supported a wide terreplein (fighting deck).

The town wall of the 1600s enclosed only the southwest and the southern periphery of the city; on the north, San Juan was open to the Atlantic, with the rocky shoreline almost the only barrier. The new project strengthened the southern wall, redesigned that of the southwest, and built a northern one so that San Juan was completely enclosed.

Nor was this all. The planning of O'Daly and Mestre reached to Boquerón Inlet. They modernized

Military engineers of the 17th and 18th centuries considered garitas *or sentry boxes absolutely "indispensable for a good defense." Strategically sited at salient points on fortress and city walls, the* garitas *afforded the sentries who occupied them unobstructed views of various parts of the town and fortifications, making it very difficult for an enemy to approach by land or sea without being discovered. A close look at San Juan's remaining* garitas *shows that they are not mass-produced copies of each other but one-of-a-kind structures that sometimes reveal details of what both the builder and the era in which they were built considered appealing in military design. Today the* garita *is known and honored as a visual symbol of Old San Juan and, by extension, of Puerto Rico.*

the works of San Antonio and San Gerónimo and built the Escambrón Battery. These made a First Line of Defense. They built a Second Line of Defense across the island midway between Boquerón Inlet and San Cristóbal.

The military installations covered more than 250 acres. Within the city walls, military boundaries left only 62 acres for public or private construction, and even this was regulated to prevent interference with military operations. Beyond the walls, the military preempted all terrain. Clashes with civilian interests were inevitable.

No one could deny that Charles III's reforms had brought progress and prosperity to San Juan. As a garrison town, most of its 3,000-plus soldiers were quartered and fed by townsfolk who earned a steady income from them. There were scores of tradespeople and a long-lasting boom in housing construction and repair. By 1776, the population totaled 6,000.

Throughout the 18th century, Britain considered Puerto Rico to be a stumbling block to trade with its rich Caribbean colonies. Rumors of impending war with the British spiced the atmosphere of San Juan, for it was known that the Englishmen were more than ever determined to have Puerto Rico, either through war or diplomacy. There was even talk that Britain would exchange Gibraltar for Puerto Rico, but this was only a diplomatic pipedream.

Fort San Gerónimo was one of three small forts built in the last quarter of the 18th century to protect the land approach to San Juan Island. It played an important role in defeating the English attack on San Juan in 1797.

The rumored war with Britain began in 1775 as the American Revolution, a struggle between Great Britain and its 13 rebellious North American colonies. The French joined the American rebels, and Spain was pulled unenthusiastically into the conflict as an ally of France. In 1779 Spanish forces began the four-year Great Siege of Gibraltar (1779-83), which had been in British hands since 1704, and planned campaigns to recover Florida. Spain sent arms and money to assist General George Washington's Continental Army. In 1781 Spanish troops rather handily took Pensacola, the capital of West Florida, and the American-French allies won at Yorktown. When the peace treaty was signed in 1783, the North Americans had gained their independence and Spain had regained Florida (which she held until 1821). Gibraltar, however, remained British.

In 1789, Spain found herself embroiled in another revolution as France erupted into violence. Louis

XVI and Marie Antoinette, his queen, tried to slip out of the country but were discovered and imprisoned. The king of Spain, Charles IV, who was related to the French monarch, tried to buy his kinsman's freedom through bribery, but failed. After Louis and Marie were executed in 1793, Spain allied herself with Austria, Prussia, and Great Britain in the war against Revolutionary France.

By 1796, the French were victorious from the Baltic to the Mediterranean. For Spain, Charles IV bought peace with the Treaty of Basel, which gave France the colony of Santo Domingo in the Caribbean. Bound ever tighter to French interests, Spain found herself aligned with France and Holland against Great Britain. Despite the loss of its former allies, Britain was almost continually at war with France from 1793 to 1815 for high stakes: a restoration of the balance of power in Europe that would allow Britain to retain control of the seas, extend its colonies, and win commercial predominance throughout the world.

Britain's ambitious objectives were threatened by the 1796 alliance of France, Spain, and Holland, giving these allies a combined naval power far superior to its own. The British were worried both about invasion of their home islands and protecting their commerce from enemy warships at sea. The British navy rose to the challenge, however, and in 1797-98 won resounding victories against the Spanish off Cape St. Vincent, Spain; against the Dutch at Camperdown, Holland; and against the French in the Battle of the Nile (Abukir Bay), Egypt.

After Spain switched to the French side in 1796, Britain moved against the Caribbean colonies of both countries. She already occupied the ports of Spanish Santo Domingo and French Guadeloupe, St. Lucia, Maria-Galante, and Port-au-Prince. Early in 1797, Admiral Sir Henry Harvey's fleet sailed to Barbados, where an army came aboard under the command of Sir Ralph Abercromby, a 62-year-old Scot and veteran of the Seven Years' War. Together they plucked Trinidad from the hands of a demoralized Spanish garrison. Emboldened by this easy success, they made ready to take the principal bastion of the Antilles—San Juan.

On 17 April 1797, the British armada appeared off the coast east of San Juan island and anchored near Cangrejos Point. There were 68 vessels with an armament of about 600 guns and manpower of about 7,000, including ships' crewmen, regulars, German auxiliaries, and French émigrés. Two of the frigates took blockading positions at the entrance to San Juan harbor.

Although San Juan was far from Europe, the tropical city was nonetheless affected by turbulence on the Continent. The British attack had been anticipated for several months and defense plans had long since been updated to take maximum advantage of the great fortifications and the patriotism of the *puertorriqueños*. Defense forces had been mobilized all over the island to aid the regular garrison. Thegovernor, don Ramón de Castro, was an able and energetic soldier who had won recognition in the capture of Pensacola from the British in 1781. With a mixed force of militia, local recruits, armed peasantry, paroled prisoners, and French privateers, the Spanish garrison almost equaled the attackers in number. Their strong fortifications mounted 376 cannon. And they were fighting for their homeland.

They did not succeed, however, in preventing Abercromby from landing 3,000 men on the beach on 18 April. His plan was to march on San Juan by land, as Cumberland had done in 1598. His men quickly took over the Cangrejos area, thus blocking San Juan's communication by land with the interior. In a polite exchange of notes, Castro courteously refused Abercromby's demand that he surrender at once and thus spare the lives of his men.

While Abercromby was setting up camps ashore (consideration for his troops was one of his virtues), Admiral Harvey, with one eye on the guns of El Morro, cautiously reconnoitered the coast west of San Juan on 20 April in search of a second landing place. Castro's men, in the meantime, had withdrawn from Cangrejos and were now improving their position at Escambrón, a strong point of the First Line of Defense. Because the British had cut the supply route to the mainland interior, canoes crossed the bay from the Cataño shore and Bayamón River to bring supplies and reinforcements to San Juan.

Abercromby's next task was to silence the batteries of San Gerónimo and San Antonio at Boquerón Inlet. Only then could he cross the inlet, take the Second Defense Line in front of San Cristóbal, and reach the major fortifications around the city. On 21

Don Ramón de Castro had been governor, intendant, and captain general of the garrison of Puerto Rico for a little more than two years when the English fleet appeared off the island's Atlantic coast east of San Juan in April 1797. An experienced soldier, he had complete confidence in his soldiers' ability to withstand the threat to the city. When General Abercromby and Admiral Harvey called upon him to surrender the city and its garrison, Governor de Castro refused, politely, vowing to resist "until I lose the last drop of my blood."

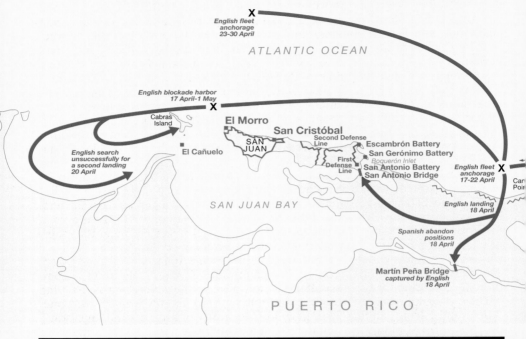

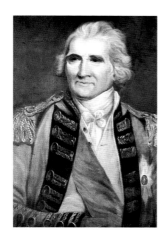

General Ralph Abercromby (top) and Admiral Sir Henry Harvey were joint commanders in the 1797 attack on San Juan. In a subsequent recounting of the Puerto Rico campaign, Abercromby all but admitted that he and Harvey had underestimated the Spanish defenders when they planned the operation, trusting too much "on the weakness of the enemy. We found him well prepared, with a garrison stronger than our force and with powerful artillery. The troops certainly were of an inferior sort, but behind walls, they could do no less than fulfill their duty with success."

April he established siege batteries and began a seven-day artillery duel with San Gerónimo and San Antonio forts—a contest that ended in the destruction of the English lines and batteries by relentless fire, not only from the two small forts and the two lines of defense but also from San Cristóbal itself.

Abercromby could not enlarge his beachhead. His landing had brought all of Puerto Rico to arms. The people rallied by the thousands. As the armed *puertorriqueños* closed in, the British position became precarious. Even as Abercromby was emplacing the siege guns, the Spanish recaptured Martín Peña Bridge, an essential gate to the interior. And on 24 April militia Sergeant Francisco Díaz raided behind British lines and brought back a number of prisoners. The British tried moving cannon onto Miraflores, a small harbor island between San Juan and the mainland, but Spanish guns stopped them.

On the 29th and 30th the Spanish counterattacked, crossing Boquerón Inlet, forcing Abercromby's men to pull back. On 1 May, as the governor was readying another attack, he learned that the English had gone, leaving behind a quantity of arms and ammunition.

Both sides suffered considerable losses in men and materiel. Abercromby's guns had badly damaged San Gerónimo and San Antonio, but he later wrote that the San Juan defenses were "both by Nature and Art, very strong," and could have withstood ten times more firepower than he had.

The final defeat of Napoleonic France at Waterloo in 1815 did not end the revolutionary changes that began with the Declaration of Independence at Philadelphia in 1776. The warfare in Europe had so engulfed Spain that her American colonies, one after another, broke away. By 1830 Puerto Rico and Cuba were the only ones that remained loyal to the mother country. France, too, except for a few small Caribbean possessions, was gone from the Americas, leaving only the Danes, the Dutch, and the British to exploit the Antilles.

With the collapse of the Spanish-American empire, San Juan was no longer important as a gateway bastion. Since Spain's rivals had achieved their goal of open ports and free trade, they no longer coveted Puerto Rico as a base of operations. San Juan's great forts were therefore downgraded from outposts of empire to local defenses.

Except for a few pirate raids on coastal towns in the early 19th century, Puerto Rico grew in peace for a hundred years after 1797. The big forts were not only maintained, but improved. After Abercromby left, battle-scarred San Gerónimo and San Antonio were rebuilt. The Second Defense Line, an earthwork, was redesigned and constructed in masonry. Later there were numerous additions and refinements to other works, such as constructing barracks and updating armament. Because the revolution in Mexico ended the subsidies that had supported the island economy for more than 200 years, Puerto Rico paid the costs of the new work after 1810.

While the fortifications were kept in good condition, San Juan was not. The population had swelled far out of proportion to the space available within the walls and the area outside was kept clear for military reasons. By 1876, 24,000 people were living inside the walls (*intramuros*) of a 62-acre city with room for only 926 buildings. Most of the houses were only one story. With no space for new housing or public buildings or even facilities for expanding manufacture and trade, the crowding was intolerable. Gradually the city's residents focused the blame for the situation on the military, for it was they who upheld the arbitrary authority of the governor. The great forts that once symbolized peace and security now seemed more like grim prison walls.

These were years of peace in Puerto Rico, but a new sense of nationalism was at work. Advocates of home rule for Puerto Rico won a few concessions, but certain Spanish governors used harsh measures to maintain control. A number of illustrious Puerto Rican patriots were imprisoned inside El Morro.

Finally in 1897, after a 40-year haggle, the island's government allowed for the expansion of the city. As *sanjuaneros* watched, the Spanish authorities demolished the Santiago Gate, which opened to the clear area east of the city. Amidst great rejoicing, the crowding population spilled over into former military zones. Eventually the entire southeastern sector of the wall, including the Santiago Ravelin and Santiago Bastion, was leveled.

No shot had been fired in anger from the great forts since 1797. Though the winds of war were blowing from Cuba, Spain's military leaders did not seem unduly concerned about Puerto Rico. In 1896, 18th-century cannon could still be seen on the ramparts. Soon afterward, a few batteries received modern guns and thicker parapets. But the work was done with a sense of futility, for there was little point in arming 18th-century fortifications against the powerful rifled guns of an armored fleet.

In support of the Cuban struggle for independence, the United States declared war on Spain in April 1898 and prepared to prevent Spanish troops and supplies from reaching Cuba. Puerto Rico was to become a theater of action. Available for its defense were about 8,000 regulars, and volunteers and partisan groups numbering about 6,000. Preparing to resist any invasion, the volunteers marched through the streets to their posts in the defense line almost every evening, after which they would march home again. As for gunnery practice, there was little either before or after war started. Several evenings a week, the citizenry gathered in the main plaza for military concerts, as was their custom.

At 5 a.m. on 12 May 1898, San Juan was jolted awake by the sound of big rifled naval guns. The troops hurried to their posts as shells burst against the fortifications and city walls. From the ramparts they saw warships about a mile offshore. It was a United States fleet commanded by Admiral William T. Sampson. As the shelling continued, terrified residents streamed out of San Juan to the countryside.

Admiral Sampson was looking for the naval squadron under Admiral Pascual Cervera y Topete, known to be en route from the Cape Verde Islands, headed presumably for San Juan. Sampson, if he did not find Cervera, intended to return to blockading Havana. He had nine warships and two smaller craft. Four warships circled northwest to El Morro, each firing a broadside as it came within 1,600-yard range. The remaining warships engaged San Cristóbal and the eastern guns of El Morro. Some came so close inshore that San Cristóbal's cannon could not be depressed enough to aim at them. Infantrymen on the north wall, however, did fire their rifles at the American vessels.

The Spanish artillery performed well. The gunners, never before in action, got off 441 rounds and scored a number of hits. Only one gun, at San Cristóbal, was put out of action. Of the 43 pieces mounted, only 28 could shoot, and these were medium-caliber pieces

San Cristóbal Bombarded

"A few minutes after five . . . in the morning [on 12 May 1898], the tremendous roar of heavy artillery made me jump from the folding cot on which I was sleeping, dressed in uniform and still wearing my small arms. I raced up to the battery of the uppermost level of the fort at a dead run. There I found a good portion of my men showing great surprise; the rest I dragged out of their sleeping quarters in short order. Since all the cannon and howitzers had been kept loaded since the 10th [when the warship USS Yale *appeared off San Juan], it was easy to open fire, <u>seven minutes after the first shot of the enemy.</u>*

"A rain of projectiles, shaking [the air] like locomotives, passed over our heads; it was a veritable storm of iron; there, at sea, where the day was beginning to lighten, could be distinguished the silhouettes of the enemy ships, illuminated from time to time by the muzzle blasts of their guns.

"I calculated the distance, at a quick look, to be 4,000 meters, and . . . shouted the command to fire at that range with bursting shell. The first primer failed, perhaps due to the inexperience or nervousness of the gunner; then I began to shoot, cannon by cannon, aiming carefully. This lasted until eight o'clock in the morning; three hours of battle against a powerful squadron; three hours that seemed to me to be three centuries."

—Captain Angel Rivero-Méndez, San Cristóbal artillery commander.

73

San Juan Under Fire 1898

Early on the morning of 12 May 1898, a month after the United States declared war on Spain, a United States naval squadron commanded by Admiral William T. Sampson (below right) arrived off San Juan. Sampson was searching for a Spanish squadron commanded by Admiral Pascual Cervera y Topete. Cervera's fleet was not at San Juan but Sampson's warships, including the USS *Iowa* pictured below, bombarded the city for almost 3 hours before resuming the search. In all, the American ships fired from 800 to 1,000 shells against the Spanish defenses. The previously untested Spanish artillerymen in the fortifications responded well during the attack but they were no match for the guns of the American ships. El Morro's walls took many hits, one of which is recorded in the photograph at right. Heavy projectiles entered the walls to a depth of 6 feet and tore large pieces out of the masonry work. One shell destroyed the upper part of the lighthouse; another pierced the Santa Bárbara Battery and lodged in the original 1540 tower. (A fragment of that shell is still embedded in the tower wall.) San Cristóbal and its outworks sustained several hits, slightly damaging the outer walls. In the city, several government buildings, the jail, a hospital, and several private residences suffered light to moderate damage during the bombardment. Spanish casualties, including civilians, were 36 dead and 20 wounded. American losses were one dead and seven wounded.

Defenders of Old San Juan

The uniforms and the weapons changed but the mission remained the same—to preserve and protect the forts of Old San Juan. From the earliest days, when soldiers wore armor and the defenses consisted only of a tower and a water battery at the harbor entrance, to 1949 and the establishment of San Juan National Historic Site, the story of these old fortifications is the story of the people who garrisoned, maintained, and defended them. (A few of these protectors are shown here and identified in the legend.) Today's guardians do not carry guns; instead they are armed with the knowledge of the role the forts played in Caribbean history and a desire to share that knowledge with the people of Puerto Rico and visitors from around the world.

1 Spanish Gunner, 1620-40
2 Spanish Infantry, 1740s
3 Freedman, Disciplined Militia of Puerto Rico, 1769
4 Spanish Infantry, 1780s
5 Officer, Puerto Rican Regiment, 1790s
6 Spanish Infantry, 1820s
7 Spanish Artillery, 1898
8 U.S. Coastal Artillery, 1910-20
9 U.S. Coastal Artillery, 1941-45
10 Ranger, National Park Service

that were not much of a threat to Sampson's five battleships and one armored cruiser.

Sampson's medium- and large-bore guns fired about a thousand rounds and pockmarked the fortification walls. Big shells tore out six-foot chunks of masonry. The top of the lighthouse at El Morro was blown off, and one shell (which still can be seen) pierced Santa Bárbara Battery and lodged in the dome of the 16th-century tower. In the city, the barracks, jail, hospital, and a number of homes were damaged. Many shells went over the city and into the waters of the bay. After three hours, having decided against a landing attempt, Sampson stopped the bombardment and sailed away. Eight weeks later, his fleet destroyed Cervera's squadron off Santiago, Cuba.

The invasion Puerto Ricans were expecting came on 25 July, but not at San Juan. General Nelson A. Miles, a 59-year-old veteran of the Indian wars in the western United States, landed troops on the southern coast at Guánica, where the opposition was only a patrol of 11 civilians, while other U.S. forces landed elsewhere on the coast. The plan was to rendezvous at San Juan and, with support from the blockading fleet, surround the city. That never happened, because on 12 August the United States and Spain negotiated a cease-fire.

At the peace talks Spain agreed to give Cuba its independence and to cede Puerto Rico, Guam, and the Philippines to the United States. The last remnants of the great empire born in 1492 were gone.

Spanish soldiers took down the bronze seal with the coat of arms that had adorned the entrance of El Morro since the 18th century. Spanish officials took this symbol of Spain's sovereignty back to the mother country. On 18 October 1898, the United States flag replaced the Spanish.

Puerto Rico now took on a new strategic importance. The outbreak of World War I showed the military value of Puerto Rico as an outpost for detecting and controlling hostile naval activity aimed at the Panama Canal or elsewhere in the Caribbean. As a result, many of the 18th-century defense structures were modified for 20th-century use. El Morro became part of the large complex of administrative, housing, and hospital units known as Fort Brooke. During World War II, coast defense observation posts and hidden command and communications centers

Three lighthouses have stood on El Morro's fifth level in its long history. The first one was constructed in 1846. A second one replaced it in 1876. The second lighthouse took a direct hit during the 1898 bombardment of El Morro by Sampson's U.S. fleet, but the brick foundation was salvaged and reused in 1899-1906 to erect the lighthouse in use here today.

were built within San Cristóbal and El Morro. These can still be seen.

For 500 years the strategic position of this island has been the primary influence on its history. Although the technology of travel and communications continues to change, the value of this *puerto rico*, this excellent harbor, remains unchanging. The island and its people are a natural bridge between the Latin American world and the English-speaking nations to the north. Puerto Rico is not just a place of sunshine and tropical beaches; it is centuries of rich Spanish history as well. The forts of Old San Juan are a part of that heritage. For the millions of visitors who explore their windswept ramparts and bastions each year, these magnificent fortifications are silent reminders of Spain's impressive and indisputable role in the history of the New World. They are also windows through which, with a little imagination, San Juan and Puerto Rico can once again become the historic gateway to the fabulous Spanish Main.

National Park Service

The National Park Service is grateful to all those persons who made the preparation and production of this handbook possible. Especial thanks is extended to San Juan National Historic Site Superintendent William P. Crawford, who drew the sugar cane and cassava plants on pages 59 and 60, and to park staff members Mark Johnson, Rosanna Weltzin, Candida Alicea, and Milagros Flores for their help in providing critical material for the book on very short notice.

This handbook was produced by members of the staff of the Division of Publications: Raymond Baker, editor; Nancy Morbeck Haack and Tom Patterson, cartographers; and Linda Meyers, designer. It is based on a book originally written by Albert Manucy and Ricardo Torres-Reyes and published by Eastern National Park and Monument Association as *Puerto Rico and the Forts of Old San Juan*. The National Park Service expresses its appreciation to the Association for its cooperation in this project. The siege drawings on page 42 are based on artwork originally developed by Albert Manucy. All illustrations not credited below are from the files of San Juan National Historic Site or of the National Park Service.

Art and History Museum of San Juan 69
The British Library, 14-15
Department of State of Puerto Rico 16
Ken Laffal cover, 4-5, 6, 8-13, 16, 38, 49, 52, 56 (photographs), 67, 78
Library of Congress 74-75 (USS *Iowa*, Sampson)
Museo Nacional Del Prado, Madrid 29, 63 (Charles III)
National Maritime Museum, Greenwich 71 (Harvey)
National Portrait Gallery, London 30, 33, 71 (Abercromby)
New York Public Library 22, 63 (O'Reilly)
Richard Schlecht 18-19, 26-27, 31, 54-55, 62-63
Arthur Shilstone 20-21, 56-57 (artwork), 64-65, 76-77